A New Conception

of

Geometry

by Academician Jingzhong Zhang

Translated by Qian Fu

Saltire Software, Inc.
Tigard, OR, USA
www.saltire.com
www.geometryexpressions.com

Published by Saltire Software 2018

ISBN: 1-882564-30-8

Saltire Software
12700 SW Hall Blvd.
Tigard, OR 97223
http://www.geometryexpressions.com/
http://www.saltire.com/
support@saltire.com

TABLE OF CONTENTS

Chapter 15

Chapter 16

Chapter 17

ACKNOWLEDGMENTS

The author acknowledges Qian Fu, Hongguang Fu and Qingxian Wang for translating this text; Lu Wang, Yuanyuan Sun and Shijie Lu for drawing the figures with the help of *Geometry Expressions* software; and editors Philip Todd and Hannah Kemper.

INTRODUCTION

JINGZHONG ZHANG

An Academician of the Chinese Academy of Sciences.

Mathematician Zhang has engaged in research into geometric algorithms and theorem proving by computer for many years. The results of his research have won him second prize in the National Technology Invention Awards, first prize of the Chinese Academy of Sciences Natural Science Award and the second prize of the National Natural Sciences Award.

Mathematician Zhang is an enthusiastic contributor in the area of mathematics education. He is engaged in reform of K-12 mathematics and calculus teaching.

He loves "Popularization of Science and Technology" (PST) projects. His "Education Mathematics " series won the Chinese Books Award. The PST works, "The Vision of the Mathematician" and others written by him, won second prize in National Scientific and Technological Progress, the Sixth National Book Award, "5-1" Achievement Award, and first prize in National PST Pioneering Awards. He edited the series, "Amusing Mathematics" which won second prize in National Scientific and Technological Progress.

Chapter 1

Keep on going

Compare the areas of two triangles

Some questions that seem very simple and common may present new ideas when given a second thought.

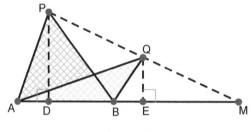

Figure 1.1

There are two triangles in Fig. 1.1: $\triangle PAB$ and $\triangle QAB$. It is obvious that the area of $\triangle PAB$ is larger than the area of $\triangle QAB$.

Further, how many times is the area of $\triangle PAB$ larger than the area of $\triangle QAB$? It cannot be figured out at first glance. Instead, we need to measure it by hand.

It is not difficult. In primary school, we have learned that the area of triangle is half the product of the base and height. First, we draw the altitudes of $\triangle PAB$ and $\triangle QAB$, PD and QE respectively, then we measure them:

$AB = 4\text{(cm)}$

$PD = 4\text{(cm)}$

$QE = 2\text{(cm)}$

Immediately, we can determine that the area of $\triangle PAB$ is 8cm² and that the area of $\triangle QAB$ is 4cm². Therefore, the area of $\triangle PAB$ is twice the area of $\triangle QAB$.

You might soon find the approach above to be an awkward one. In fact, it is not necessary at all to calculate the area of the two triangles because they have one common side *AB*, which can be taken as their common base. Two triangles with a common base are called **Co-Base Triangles**. The ratio of their areas equals the ratio of their heights, hence:

$$\frac{\triangle PAB}{\triangle QAB} = \frac{PD}{QE} = \frac{4}{2} = 2$$

(1.1)

That is, the area of ΔPAB is twice the area of ΔQAB.

In equation (1.1), for the sake of convenience in writing, we use ΔPAB to denote its area. Such notation will be used throughout this book. In this book, ΔPAB sometimes denotes the triangle PAB, and sometimes it denotes the area of triangle PAB. It doesn't matter since we can infer the meaning from the context.

We can calculate the ratio of the area of ΔPAB and ΔQAB just by measuring the altitudes, without measuring the base. This method makes full use of the condition that they have a common base. Apparently, it is smarter than calculating their areas directly. Can we come up with an even better approach?

To measure the altitude, one must first use the right angle set square to draw two altitudes, then measure them by ruler. Are there any simpler approaches?

Indeed, there are. Let's look at Fig. 1-1. Suppose that line *AB* intersects line *PQ* at point *M*, and we can measure the length of line segment *PM* and *QM*. When we measure the two line segments, it is not necessary to draw the perpendicular lines or to measure them. By measuring *PM*, we have *PM* = 8cm, *QM* = 4cm, which leads to the same result:

$$\frac{\triangle PAB}{\triangle QAB} = \frac{PM}{QM} = \frac{8}{4} = 2$$

(1.2)

Why does this make sense?

Readers who have learned similar triangles will find that ΔPDM ~ ΔQEM, thus:

$$\frac{PD}{QE} = \frac{PM}{QM} \tag{1.3}$$

which means that given the ratio of PM and QM, we know the ratio of PD and QE, and consequently we know the ratio of ΔPAB and ΔQAB.

Wonderful! You have found a better method, and, in addition, elaborated on the reasons using your knowledge of similar triangles. Congratulations!

But you should not feel content yet. You should ask whether there are any simpler ways to derive the following equation.

$$\frac{\triangle PAB}{\triangle QAB} = \frac{PM}{QM} \tag{1.4}$$

For example, can you explain the reason why equation (1.4) holds, to a pupil who doesn't yet know about similar triangles?

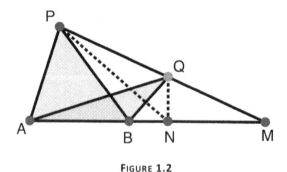

FIGURE 1.2

There is a method. Take a point N on AB such that MN = AB, as shown in Fig. 1-2. Then

$$\triangle PAB = \triangle PMN$$

$$\triangle QAB = \triangle QMN$$

So
$$\frac{\triangle PAB}{\triangle QAB} = \frac{\triangle PMN}{\triangle QMN} = \frac{PM}{QM} \tag{1.5}$$

Here, we used the fact that the ratio of areas of **Co-Altitude Triangles** (two triangles with the same altitude and different bases) equals the ratio of their bases. ΔPMN and ΔQMN are co-altitude triangles since PM and QM can be taken as the respective base of ΔPMN and ΔQMN. Their common altitude is not drawn in the Fig.1.2.

To get equation (1.4), we take a point *N* on line *AB* and connect the line segment *PN* and *QN*. If you don't want to add those auxiliary points and auxiliary lines, you can use those existing co-altitude triangles as transitions:

$$\frac{\triangle PAB}{\triangle QAB} = \frac{\triangle PAB}{\triangle PBM} \cdot \frac{\triangle PBM}{\triangle QBM} \cdot \frac{\triangle QBM}{\triangle QAB} = \frac{AB}{BM} \cdot \frac{PM}{QM} \cdot \frac{BM}{AB} = \frac{PM}{QM} \qquad (1.6)$$

Equation 1.4 was also derived in this way, yet without auxiliary points and lines.

Now, examining our thinking process, we can obtain some beneficial enlightenment.

First, don't neglect those questions that look plain and simple. There may be something you haven't understood that has been left behind.

Second, just give a second thought to whether there's a smarter solution to the problem which you have just solved.

Third, more consideration may refine the analysis of a better method. Can you, for example, explain the method with less knowledge assumed or with knowledge which is more elementary?

It's not over. You may learn by analogy. Δ*PAB* and Δ*QAB* in Fig. 1.1 have one common side, *AB*. But their positions are specific to the diagram.

Does equation (1.4) hold for any pair of triangles with a common side, no matter what configuration?

Analyzing questions and proposing questions in this way, we have already extracted the general concept - "two triangles with a common side" from the special Δ*PAB* and Δ*QAB* in Fig. 1.1. And with the general concept, we can suggest more general questions and find more general laws.

ADDITIONAL PROBLEMS

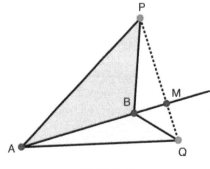

FIGURE 1.3

[P1.1] In the question mentioned at the beginning, if points *P* and *Q* lie on the different sides of line *AB*, the ratio of ΔPAB and ΔQAB can be converted into the ratio of which two line segments? (As shown in Fig.1.3)

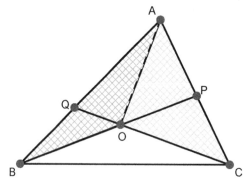

FIGURE 1.4

[P1.2] Given *AP:PC* = 4:3, *AQ:QB* = 3:2, what is the ratio of ΔAOB to ΔAOC? (As shown in Fig.1.4)

CHAPTER 2

LEARN BY ANALOGY

Co-Side Theorem and Its Applications

Two triangles with a common side are called **Co-Side Triangles.**

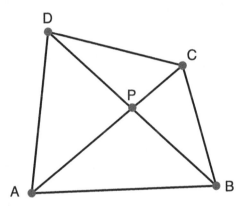

FIGURE 2.1

There are congruent triangles and similar triangles in your geometry textbook, but no co-side triangles. In fact, co-side triangles have more chance of appearing in a geometric figure. For example, we take any four points in the plane and label A, B, C and D, as shown in Fig. 2.1. There are generally no congruent triangles or similar triangles. But you can find many co-side triangles. (if you count, you can find that there are 6 pairs of co-side triangles without considering the intersection point P. If P is considered, there are 18 pairs!)

Now, let's study co-side triangles. Are there any general laws for co-side triangles which are worth mentioning?

In the previous chapter we discussed a pair of special co-side triangles $\triangle PAB$ and $\triangle QAB$, we discovered Equation (1.4), i.e. the ratio of areas of co-side triangles can be transformed into a ratio of two line segments, which can be stated as follows:

Co-Side Theorem: Suppose that line AB intersects line PQ at point M *(Fig.2.2)*, then

$$\frac{\triangle PAB}{\triangle QAB} = \frac{PM}{QM}$$

(2.1)

Proof: There are 4 cases:

(1) Points P and Q are on different sides of line AB, and points A and B are on different sides of line PQ;

(2) Points P and Q are on same side of line AB, Points A and B are on different sides of line PQ;

(3) Points P and Q are on different sides of line AB, and points A and B are on same side of line PQ;

(4) Points P and Q are on same side of line AB, and points A and B are on same side of line AB.

The following will work on any one of them. Just assume that points M and B don't coincide. (Otherwise, the equation obviously holds).

$$\frac{\triangle PAB}{\triangle QAB} = \frac{\triangle PAB}{\triangle PMB} \cdot \frac{\triangle PMB}{\triangle QMB} \cdot \frac{\triangle QMB}{\triangle QAB}$$

$$= \frac{AB}{MB} \cdot \frac{PM}{QM} \cdot \frac{MB}{AB} = \frac{PM}{QM} \qquad \square$$

Here the symbol "$\square$" denotes that the deduction or the calculation process is over (denoted in other texts as "*QED*").

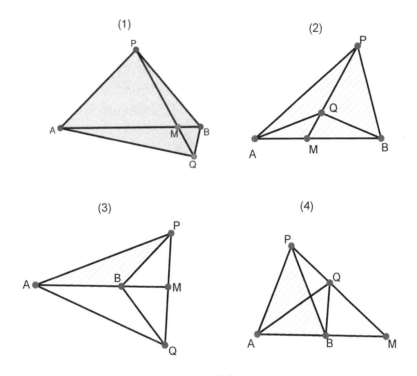

FIGURE 2.2

There is another method of proof which takes a point N on line AB such that $MN = AB$. We have immediately:

$$\frac{\triangle PAB}{\triangle QAB} = \frac{\triangle PMN}{\triangle QMN} = \frac{PM}{QM}$$

Comparing to the practice of the previous chapter, we discover that some methods of solving special problems still work well for more general problems. While deducing Equation (1.4), this method worked well applied to Fig.1-1. But we find that the method is also suitable for the four cases of Fig.2.2.

After proving the theorem, let's summarize the experience and extract general strategies which may be applied to new problems. In the second of the above two methods, it is ingenious to use auxiliary points to simplify this problem. But the first method is also very useful, which is called the **_Transition Method or Bridging Method_**. Since we can't find the ratio of $\triangle PAB$ to $\triangle QAB$ easily, we utilize other triangles as bridges to transit gradually. In fact, it is easy to find ratios of $\triangle PAB$ to $\triangle PMB$, $\triangle PMB$ to

$\triangle QMB$, $\triangle QMB$ to $\triangle QAB$. Now the transition is done. Sometimes, this method can solve very difficult problems. In this book, it will be used repeatedly.

The more useful a particular theorem is, the more important it is. Here, we give several examples to show the importance of the Co-Side Theorem.

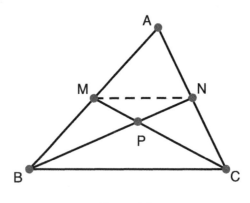

FIGURE 2.3

[Ex. 2.1] In Fig. 2.3, M and N are midpoints of the sides AB and AC respectively. Line segment BN intersects line segment CM at point P.

Prove: $CP = 2PM$.

$$\frac{CP}{PM} = \frac{\triangle CBN}{\triangle MBN}$$

Analysis: The question is to calculate the ratio of CP to PM. The ratio of line segments can be transformed into the ratio of areas by the Co-Side Theorem:

Because N and M are the midpoints of AC and AB respectively, $\triangle CBN$ is half of $\triangle ABC$ and $\triangle MBN$ is half of $\triangle ABN$. Thus, $\triangle MBN$ is a quarter of $\triangle ABC$.

This illuminates the path to a solution. Based on this analysis, we can write down the deduction.

Proof: According to Co-Side Theorem, we have

$$\frac{CP}{PM} = \frac{\triangle CBN}{\triangle MBN} = \frac{\triangle CBN}{\triangle ABN} \cdot \frac{\triangle ABN}{\triangle MBN} = \frac{CN}{AN} \cdot \frac{AB}{MB} = 2 \qquad \square$$

Ex. 2.1 is a well-known theorem in plane geometry: The distance from the **Centroid of a Triangle** to the vertex is twice the distance from the centroid to the midpoint on the opposite side (In Fig. 2.3, P is the centroid of $\triangle ABC$). The more general situation is:

[Ex. 2.2] In Ex. 2.1, if $AM = \lambda MB$ and $AN = \mu NC$

Determine the ratio: $\dfrac{CP}{PM}$

Solution: According to Co-Side Theorem, we have:

$$\frac{CP}{PM} = \frac{\triangle CBN}{\triangle MBN} = \frac{\triangle CBN}{\triangle ABN} \cdot \frac{\triangle ABN}{\triangle MBN} = \frac{CN}{AN} \cdot \frac{AB}{MB} = \frac{1+\lambda}{\mu} \qquad \square$$

By the same method, we can solve some interesting area calculation problems, for example:

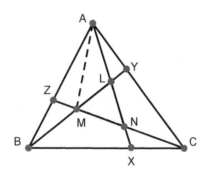

Figure 2.4

[Ex. 2.3] In **$\triangle ABC$**, take points X, Y, Z on its three sides BC, CA and AB respectively such that $CX = \frac{1}{3}BC$, $AY = \frac{1}{3}AC$ and $BZ = \frac{1}{3}AB$. The intersection points of segments AX, BY and CZ are the vertices of $\triangle LMN$, as shown in Fig. 2.4. Determine the ratio of $\triangle LMN$ to $\triangle ABC$.

Solution: By the given conditions, first we calculate the ratios of $\triangle MBC$, $\triangle NCA$, and $\triangle LBA$ to $\triangle ABC$ respectively.

$$\frac{\triangle ABC}{\triangle MBC} = \frac{\triangle ABM + \triangle BCM + \triangle ACM}{\triangle MBC} = \frac{AY}{CY} + 1 + \frac{AZ}{BZ} = \frac{1}{2} + 1 + 2 = \frac{7}{2}$$

$$\therefore \quad \triangle MBC = \frac{2}{7} \triangle ABC$$

Likewise,

$$\triangle LAB = \frac{2}{7} \triangle ABC$$

$$\triangle NCA = \frac{2}{7} \triangle ABC$$

$$\therefore \quad \triangle LMN = \frac{1}{7} \triangle ABC \qquad \square$$

The three examples are relatively simple. In the following, we will see some very hard problems which can be solved easily and beautifully by the Co-Side Theorem!

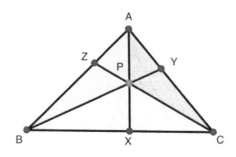

FIGURE 2.5

[Ex. 2.4] Take any point *P* in the interior of △*ABC*. Connect *AP, BP* and *CP* such that lines *AP, BP, CP* intersect *BC, AC, AB* at points *X, Y, Z*, respectively.

Prove: $\qquad \dfrac{PX}{AX} + \dfrac{PY}{BY} + \dfrac{PZ}{CZ} = 1$

Proof: As shown in Fig. 2.5, by the Co-Side Theorem, we have

$$\frac{PX}{AX} + \frac{PY}{BY} + \frac{PZ}{CZ} = \frac{\triangle PBC}{\triangle ABC} + \frac{\triangle PAC}{\triangle ABC} + \frac{\triangle PAB}{\triangle ABC} = \frac{ABC}{ABC} = 1 \qquad \square$$

12

[Ex. 2.5] From Fig.2.5,

Prove:
$$\frac{AZ}{BZ} \cdot \frac{BX}{XC} \cdot \frac{CY}{YA} = 1$$

Proof: According to Co-Side Theorem, we have

$$\frac{AZ}{ZB} \cdot \frac{BX}{XC} \cdot \frac{CY}{YA} = \frac{\triangle PAC}{\triangle PBC} \cdot \frac{\triangle PAB}{\triangle PAC} \cdot \frac{\triangle PBC}{\triangle PAB} = 1 \qquad \square$$

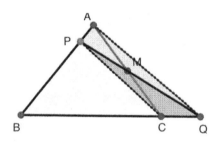

FIGURE 2.6

[Ex, 2.6] (Beijing High School Student Mathematics Competition,China,1978) In $\triangle ABC$, M is the midpoint of AC. Draw a line through point M such that its end point P lies on AB and its endpoint Q lies on the extension of line BC.

Prove:
$$\frac{PA}{PB} = \frac{QC}{QB}$$

Proof:
$$\frac{PA}{PB} = \frac{\triangle QPA}{\triangle QPB} = \frac{\triangle QPA}{\triangle QPC} \cdot \frac{\triangle QPC}{\triangle QPB} = \frac{AM}{MC} \cdot \frac{QC}{QB} = \frac{QC}{QB} \qquad \square$$

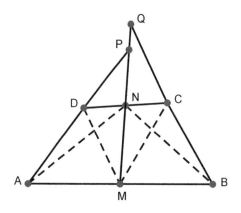

FIGURE 2.7

[Ex. 2.7] In quadrilateral *ABCD*, *AD* = *BC*. *M*, *N* are the midpoints of *AB* and *CD*, respectively. Extend *AD* and *BC* to intersect the extension of *MN* at points *P* and *Q* respectively. As shown in Fig. 2.7.

Prove: $PD = QC.$

Proof: According to Co-Side Theorem and the given conditions, we have

$$\frac{PA}{PD} = \frac{\triangle AMN}{\triangle DMN} = \frac{\triangle BMN}{\triangle CMN} = \frac{QB}{QC}$$

Plugging *PA=PD+DA* and *QB=QC+CB* into the equation above, with the condition *DA=CB*, we have *PD=QC*.

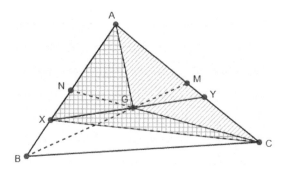

FIGURE 2.8

[Ex. 2.8] (Mathematics Competition of Anhui Province, China,1979) Point *G* is the centroid of △*ABC*. (It is the intersection point of three medians of the triangle.) Draw line XY through point *G* such that X and Y lie on *AB, AC* respectively.

14

Prove: $GX \leq 2GY$

Proof: As shown in Fig.2.8, we have

$$\frac{GY}{XY} = \frac{\triangle GAC}{\triangle XAC} \geq \frac{\triangle GAC}{\triangle ABC} = \frac{1}{3}$$

which shows that $GX \leq \frac{2}{3}XY$, thus $GX \leq 2GY$.

In the proof, we used the equation: $\triangle GAC = \frac{1}{3}\triangle ABC$. This is because M and N are the midpoints of AC and AB respectively. Thus, we have

$$\triangle GAB = \triangle GBC \text{ and } \triangle GBC = \triangle GAC$$

These examples are enough for now. If you are interested in the methods introduced here, try the exercises below. If you're still unable to solve them after a great effort, you may refer to the hints and keys in chapter 17.

ADDITIONAL PROBLEMS

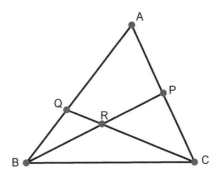

FIGURE 2.9

[P2.1] Given: P is the midpoint of AC. Point Q lies on AB and $AQ = 2QB$, as shown in Fig. 2.9.

Determine: $\quad \dfrac{PR}{BR}, \dfrac{QR}{CR}, \dfrac{\triangle RBC}{\triangle ABC}$

[P2.2] In $\triangle ABC$, take point X, Y, Z on side BC, CA and AB respectively, such that $BX = XC$, $CY = 2YA$ and $AZ = 3ZB$. What is the ratio of the area of the triangle bounded by the lines AX, BY and CZ to the area of $\triangle ABC$?

[P2.3] (Hungary Mathematics Competition, 1943) Take a point P in the interior of $\triangle ABC$.

Prove: The maximum distance from P to one of $\triangle ABC's$ vertices, is at least twice the minimum.

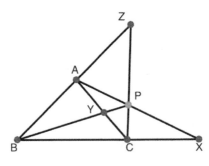

FIGURE 2.10

[P2.4] Take a point P which is outside of $\triangle ABC$ and inside $\angle ABC$. The lines AP, BP and CP intersect the sides BC, CA and AB at points X, Y and Z respectively, as shown in Fig. 2.10.

Prove:
$$\frac{PX}{AX} + \frac{PZ}{CZ} - \frac{PY}{BY} = 1$$

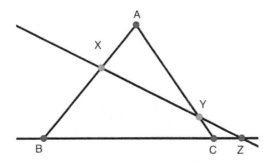

FIGURE 2 - 11

[P2.5] Take points X and Y on side AB and AC, respectively. The line XY intersects the extension of BC at point Z, as shown in Fig. 2.11.

Prove:
$$\frac{AX}{XB} \cdot \frac{BZ}{ZC} \cdot \frac{CY}{YA} = 1$$

[P2.6] From Fig.2.7 we change the conditions: points M and N are <u>not</u> the midpoints of AB and CD respectively; $AD \neq BC$.

Let:
$$\frac{PD}{AD} = \frac{QC}{BC}$$

Prove:
$$\frac{DN}{CN} = \frac{AM}{BM}$$

[P2.7] Take a point P on side BC of $\triangle ABC$. Draw a line through P which intersects AC at point X and the extension of side AB at point Y.

Prove:
$$\frac{PY}{PX} > \frac{PB}{PC}$$

CHAPTER 3

THINK FROM THE CONTRARY

Co-Side Triangles and Parallel Lines

We introduced the Co-Side Theorem with the **premise** that line *AB* intersects line *PQ* at point *M*. But what if the two lines have no intersection point?

The way of proposing such questions like the above is called ***thinking from the contrary***. For many propositions in math, thinking about them from the contrary can open up a new window.

Is it possible that the two lines have no intersection point? Of course, when $\triangle PAB$ equals $\triangle QAB$, and points *P*, *Q* are on the same side of line *AB*, they don't intersect with each other. Please look at the following examples:

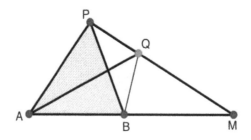

FIGURE 3.1

[Ex. 3.1] Prove: If points *P* and *Q* are on the same side of line *AB*, and line *PQ* doesn't intersect *AB*, then *PQ* ∥ *AB*.

Proof: (proof by contradiction) Suppose the extension line of *PQ* intersects the line *AB* at point *M*, as shown in Fig. 3.1. Then

$$\frac{\triangle PAB}{\triangle QAB} = \frac{PM}{QM} > 1$$

That is, $\triangle PAB > \triangle QAB$. This contradicts $\triangle PAB = \triangle QAB$. So this is a contradiction, and the assumption is false.

Conversely, if $PQ \parallel AB$, then $\triangle PAB = \triangle QAB$, which can be proved in the same way.

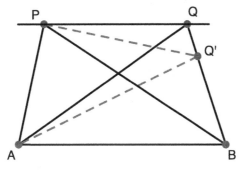

FIGURE 3.2

[Ex. 3.2] Given: $PQ \parallel AB$,

Prove: $\triangle PAB = \triangle QAB$.

Proof: (proof by contradiction) If $\triangle PAB \neq \triangle QAB$, then we can take another point Q' on segment BQ such that $\triangle Q'AB = \triangle PAB$ (as shown in Fig.3.2). Then $PQ' \parallel AB$ holds by the result of Ex. 3.1. Hence, there are two lines through point P that are parallel to line AB. This is a contradiction.

Thus, co-side triangles and parallel lines are related and this relationship is very useful in a lot of cases.

 [Ex. 3.3] Given: parallelogram $ABCD$, the diagonal AC intersects the diagonal BD at point O.

Prove: $AO = CO$ (as shown in Fig. 3.3).

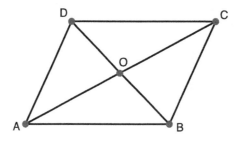

FIGURE 3.3

Proof:
$$\frac{AO}{CO} = \frac{\triangle ABD}{\triangle BCD} = \frac{\triangle ABC}{\triangle ABC} = 1$$

Do you know why? First, apply Co-Side Theorem. Then $\triangle ABD = \triangle ABC$ since $DC \parallel AB$, and $\triangle ABC = \triangle BCD$ since $AD \parallel BC$. This is just the desired result.

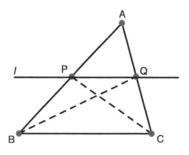

FIGURE 3.4

[Ex. 3.4] As shown in Fig. 3.4, line l is parallel to line BC, and it intersects sides AB and AC of $\triangle ABC$ at P and Q respectively.

Prove:
$$\frac{AP}{PB} = \frac{AQ}{QC}$$

Proof:
$$\frac{AP}{PB} = \frac{\triangle APQ}{\triangle BPQ} = \frac{\triangle APQ}{\triangle CPQ} = \frac{AQ}{QC}$$

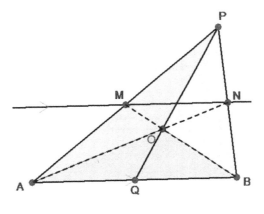

FIGURE 3.5

[Ex. 3.5] (National Middle School Mathematics Competition, China,1978) **Given**: Line segment *AB* is parallel to line *l*. Take point *P* which is neither on *AB* nor *l*. Draw lines *PA* and *PB* such that they intersect *l* at points *M* and *N*, respectively. Connect *AN* and *BM* such that *AN* intersects *BM* at point *O*. Connect *PO* to *AB* at point *Q*. As shown in Fig. 3.5.

Prove: $AQ = BQ$.

There is a story about this problem. When the *National Middle School Mathematics Competition* was held in 1978, Professor Luogeng Hua, a master of mathematics, was in charge of the examination board in Beijing. Professor Buqing Su, the famous mathematician, wrote to Professor Hua and suggested that he consider the following problem: given points *A*, *B* and line *l* on the plane, where line *l* is parallel to line segment *AB*, find the midpoint of line segment *AB* using only a ruler.

It was the examination board's view that this question was a little difficult. So it was revised into Ex. 3.5: the question is how to find the midpoint and prove it.

It is not difficult to prove by the properties of co-side triangles. According to the Co-Side Theorem, we have the following proof.

Proof:

$$\frac{AQ}{BQ} = \frac{\triangle\,AOP}{\triangle\,BOP} = \frac{\triangle\,AOP}{\triangle\,AOB}\cdot\frac{\triangle\,AOB}{\triangle\,BOP} = \frac{PN}{BN}\cdot\frac{AM}{PM}$$

$$= \frac{\triangle\,PMN}{\triangle\,BMN}\cdot\frac{\triangle\,AMN}{\triangle\,PMN} = \frac{\triangle\,AMN}{\triangle\,BMN} = 1 \qquad \square$$

$MN \parallel AB$ is used in the last step.

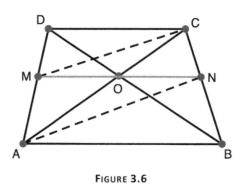

FIGURE 3.6

[Ex. 3.6] In trapezoid $ABCD$, two diagonals intersect at point O. Draw a line through O which is parallel to AB such that it intersects AD and BC at M and N respectively.

Prove: $MO = NO.$

Proof:

$$\frac{MO}{NO} = \frac{\triangle\,MAC}{\triangle\,NAC} = \frac{\triangle\,AOD}{\triangle\,BOC} = \frac{\triangle\,ABD - \triangle\,ABO}{\triangle\,ABC - \triangle\,ABO} = 1 \qquad \square$$

In this proof, the key is to figure out that $\triangle MAC = \triangle AOD$ and $\triangle NAC = \triangle BOC$, where the area segmentation techniques are used.

$$\triangle MAC = \triangle MAO + \triangle MOC = \triangle MAO + \triangle MOD = \triangle AOD$$

$$\triangle NAC = \triangle CON + \triangle AON = \triangle CON + \triangle BON = \triangle BOC$$

The area segmentation techniques will be more thoroughly exploited in the following examples.

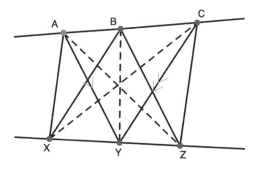

FIGURE 3.7

[Ex. 3.7] Given: points A, B and C are collinear, as are points X, Y and Z.

$AY \parallel BZ$, $BX \parallel CY$.

Prove: $AX \parallel CZ$ (as shown in Fig. 3.7).

Analysis: It's sufficient to prove $\triangle AXC = \triangle AXZ$. According to Fig. 3.7, we have

$$\triangle AXC = \triangle AXB + BXC = AXB + BXY = S_{ABXY}$$

(S_{ABXY} denotes the area of the quadrilateral $ABXY$. This notation will be used throughout the book.)

$$\triangle AXZ = \triangle AXY + AYZ = AXY + BAY = S_{ABXY}$$

$$\therefore \quad \triangle AXC = \triangle AXZ$$

ADDITIONAL PROBLEMS

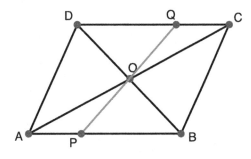

FIGURE 3.8

[P3.1] In parallelogram $ABCD$, the two diagonals intersect at point O. Draw a line through point O such that it intersects AB and CD at points P and Q respectively (as shown in Fig. 3.8).

Prove: $PO = QO$.

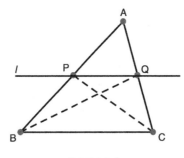

FIGURE 3.9

[P3.2] In Fig. 3.9 given that $\dfrac{PA}{BA} = \dfrac{QA}{CA}$,

Prove: $PQ \parallel BC$.

[P3.3] In Fig. 3.5, prove: line PO also bisects line segment MN.

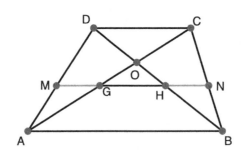

FIGURE 3.10

[P3.4] As an extension of Ex. 3.6, if $MN \parallel AB$ and MN doesn't go through point O. MN intersects the two diagonals at point G and H, respectively (as shown in Fig. 3.10).

Prove: $MG = NH$.

CHAPTER 4

FIELD PROBLEM AND POINT OF DIVISION FORMULA

There is a field which is an irregular quadrilateral *ABCD*, as shown in Fig. 4.1. Take the trisection points of each side and connect these trisection points on the opposite side. Then, the figure comes into a Chinese character "井". This field has been cut into nine pieces by the character " 井". The area of each piece is not equal because of the irregularity of the quadrilateral. But it happenes to be that whatever the field is, the area of the center piece is one-ninth of the whole area!

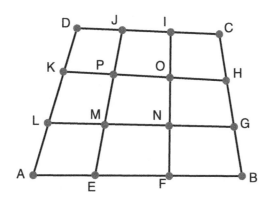

FIGURE 4.1

However, it's very hard to prove this interesting inference.

When encountering a hard but interesting problem, you are determined to tackle it. But what if it's too tough? What can you do? Here is a very useful rule:

"If the present problem is too hard, you can instead solve a similar problem that is easier."

Let's avoid this difficult problem and consider a simpler one. As shown in Fig. 4.1, can we prove the area of quadrilateral *KLGH* is one-third of that of *ABCD*?

This is not difficult:

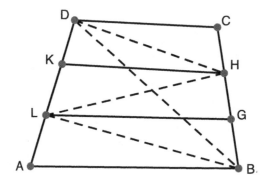

FIGURE **4.2**

[Ex4.1] In Fig. 4.1, points *E, F, G, H, I, J, K* and *L* are trisection points of sides *AB, BC, CD* and *DA, respectively.*

Prove: $S_{ABCD} = 3S_{KLGH} = 3\,S_{EFIJ}$.

Proof: As shown in Fig4.2.

$$\triangle ABL = \frac{1}{3} \triangle ABD$$

$$\triangle CDH = \frac{1}{3} \triangle CDB$$

Adding these two equations, we get

$$\triangle ABL + \triangle CDH = \frac{1}{3} S_{ABCD}$$

Hence

$$S_{KLGH} = \triangle LGH + \triangle LKH = \frac{1}{2}\left(\triangle LBH + \triangle LHD\right)$$

$$= \frac{1}{2}\left(S_{ABCD} - \triangle ABL - \triangle CDH\right)$$

26

$$= \frac{1}{2}\left(S_{ABCD} - \frac{1}{3}S_{ABCD}\right) = \frac{1}{3}S_{ABCD}$$

Likewise, we have

$$S_{EFIJ} = \frac{1}{3}S_{ABCD}$$

The easier problem has been finished. Now, if we can prove the area of *MNOP* is one third of *KLGH*, then the original problem will be solved.

Before proving this problem, let's think about the following question.

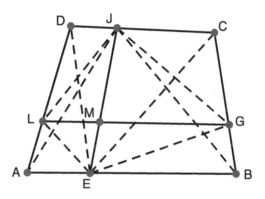

FIGURE 4.3

[Ex4.2] **Given:** The same conditions as of Ex4.1.

Prove: Points *M* and *N* are trisection points of *LG*; *P* and *O* are trisection points of *KH*.

Proof: As shown in Fig. 4.3, what needs proving is

$$MG = 2ML$$

For this purpose, we only need to prove:

$$\triangle GEJ = 2 \triangle LEJ$$

Through a detailed analysis, we have:

$$\triangle GEJ = S_{BEJC} - \triangle BEG - \triangle CJG = S_{BEJC} - \frac{1}{3}\triangle BEC - \frac{2}{3}\triangle BJC$$

$$= S_{BEJC} - \frac{1}{3}\left(S_{BEJC} - \triangle CEJ\right) - \frac{2}{3}\left(S_{BEJC} - \triangle BEJ\right)$$

$$= \frac{1}{3}\triangle CEJ + \frac{2}{3}\triangle BEJ$$

$$= 2\left(\frac{1}{3}\triangle DEJ + \frac{2}{3}\triangle AEJ\right)$$

$$\triangle LEJ = S_{AEJD} - \triangle AEL - \triangle DJL = S_{AEJD} - \frac{1}{3}\triangle AED - \frac{2}{3}\triangle DJA$$

$$= S_{AEJD} - \frac{1}{3}\left(S_{AEJD} - \triangle DEJ\right) - \frac{2}{3}\left(S_{AEJD} - \triangle AEJ\right)$$

$$= \frac{1}{3}\triangle DEJ + \frac{2}{3}\triangle AEJ$$

which shows ΔGEJ=2ΔLEJ. So, MG =2ML, i.e. point M is a trisection point of LG. Likewise, N is the trisection point of LG; P and O are the trisection points of KH.

Thus, the difficult problem has been solved by breaking up it into parts.

Before the problem is solved, you may begin with a similar but simpler one. After solving it, you should think about it further. About what? Here are some aspects that we should consider:

(1) Can we deal with harder and more general problems?

(2) For the same problem, can we deal with it by a simpler and better method?

(3) Can we summarize these methods and apply them to solving other problems?

It is easy to propose a harder and more general problem. For example, trisection points can be replaced with points dividing equally into five parts, or, trisection points on the horizontal and points dividing equally into five parts on the vertical. How about points dividing equally into four parts and points dividing equally into six parts? Moreover, can we calculate the area of sections other than the center piece? If not, what conditions should be given here? Such questions and more all deserve careful consideration.

It is not easy to solve a problem simply and elegantly, since it's always difficult to find a starting point. If unable to come up with a solution at the moment, you should mull it over patiently instead of urgently producing an answer.

What counts most is to summarize and understand the application of these methods. With the help of them, you will no longer be afraid of hard problems and more likely to come up with good approaches.

Analyzing the solution of Ex4.2, you will notice that the key to this problem lies in calculating the area of $\triangle GEJ$ and $\triangle LEJ$. When calculating, we're assumed to know $\triangle CEJ$, $\triangle BEJ$ and the position of G on BC. Thus, the area of $\triangle GEJ$ can be determined. If $\triangle DEJ$, $\triangle AEJ$ and the position of L on side AD are known, then the area of $\triangle LEJ$ can also be determined. Summarizing this method, we can get a very useful formula:

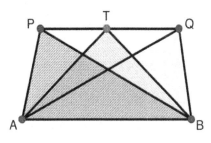

FIGURE 4.4

Point Of Division Formula: As shown in Fig. 4.4, suppose line segment PQ doesn't intersect line AB, and that point T is on PQ and $PT=\lambda PQ$, then

$$\triangle \textbf{TAB} = \lambda \triangle \textbf{QAB} + (1 - \lambda) \triangle \textbf{PAB} \qquad (4.1)$$

Proof 1: Let S denote the area of quadrilateral $ABQP$, then

$$
\begin{aligned}
\triangle TAB &= S - \triangle PAT - \triangle QBT \\
&= S - \lambda \triangle PAQ - (1 - \lambda) \triangle PBQ \\
&= S - \lambda(S - \triangle QAB) - (1 - \lambda)(S - \triangle PAB) \\
&= S - \lambda S - (1 - \lambda)S + \lambda \triangle QAB + (1 - \lambda) \triangle PAB \\
&= \lambda \triangle QAB + (1 - \lambda) \triangle PAB
\end{aligned}
$$

The proof above is a replicate of Ex4.2, which does not apply to all situations.

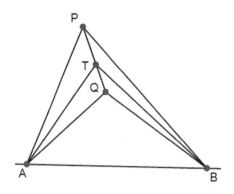

FIGURE 4.5

As shown in Fig. 4.5, the proof should be modified to the following:

$$\triangle TAB = S - \triangle PAT + \triangle QBT$$
$$= S - \lambda \triangle PAQ + (1 - \lambda) \triangle PBQ$$
$$= S - \lambda(S - \triangle QAB) + (1 - \lambda)(\triangle PBQ - S)$$
$$= \lambda \triangle QAB + (1 - \lambda) \triangle PAB$$

Applying the Co-Side Theorem, we can give a unified proof.

Proof 2: If $PQ \parallel AB$, then $\triangle TAB = \triangle QAB = \triangle PAB$, and the formula holds trivially.

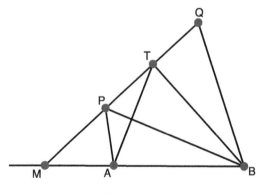

FIGURE 4.6

Hence, suppose AB intersects PQ at point M (as shown in Fig. 4.6). According to the

Co-Side Theorem we have: $\qquad \dfrac{\triangle PAB}{PM} = \dfrac{\triangle TAB}{TM} = \dfrac{\triangle QAB}{QM}$

consequently
$$\frac{\triangle TAB - \triangle PAB}{TM - PM} = \frac{\triangle QAB - \triangle TAB}{QM - TM}$$

namely
$$\frac{\triangle TAB - \triangle PAB}{\triangle QAB - \triangle TAB} = \frac{TM - PM}{QM - TM} = \frac{PT}{TQ} = \frac{\lambda}{1 - \lambda}$$

that is
$$(1 - \lambda)(\triangle PAB - \triangle TAB) = \lambda(\triangle TAB - \triangle QAB)$$

solving for $\triangle TAB$:
$$\triangle TAB = (1 - \lambda)\triangle PAB - \lambda \triangle QAB$$

With the point of division formula, solving the **Field Problem** proposed previously is no longer difficult at all. We even have a more general method, such as the following.

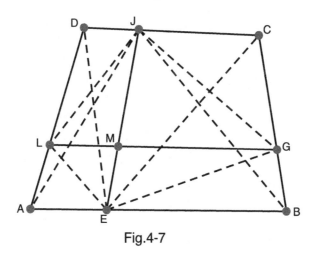

Fig.4-7

[Ex4.3] In quadrilateral *ABCD*, take points *E*, *G*, *J* and *L* on sides *AB*, *BC*, *CD* and *DA*, respectively (as shown in Fig. 4.7). Connect *LG* and *EJ* such that they intersect at *M*.

Given:
$$\frac{AE}{AB} = \frac{DJ}{DC} = \lambda$$

Prove:
$$\frac{EM}{EJ} = \mu, \qquad \frac{LM}{LG} = \lambda$$

Proof: Applying the Co-Side Theorem and point of division formula, we have

$$\frac{LM}{GM} = \frac{\triangle LEJ}{\triangle GEJ} = \frac{\mu \triangle DEJ + (1-\mu) \triangle AEJ}{\mu \triangle CEJ + (1-\mu) \triangle BEJ}$$

$$= \frac{\mu(\lambda \triangle DEC) + (1-\mu)(\lambda \triangle AJB)}{\mu(1-\lambda) \triangle DEC + (1-\mu)(1-\lambda) \triangle AJB}$$

$$= \frac{\lambda}{1-\lambda} \cdot \frac{\mu \triangle DEC + (1-\mu) \triangle AJB}{\mu \triangle DEC + (1-\mu) \triangle AJB} = \frac{\lambda}{1-\lambda}$$

Thus $\qquad \dfrac{LM}{LG} = \lambda$.

Likewise $\qquad \dfrac{EM}{EJ} = \mu$.

With the result of Ex4.3, we can immediately see that points M and N are the trisection points of LG and P and O are trisection points of KH in Fig. 4.1. From this, the **Field Problem** can be easily solved.

In the point of division formula, there is a condition that line segment PQ doesn't intersect line AB. If it does, the formula doesn't hold. As shown in Fig. 4.8, when point T is approaching point M, which is the intersection point of PQ and AB, the area of $\triangle TAB$ can approach zero. So formula (1) doesn't hold any more.

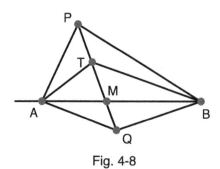

Fig. 4-8

For this case, there is another formula.

Supplement of Point of Division Formula: Suppose line segment PQ intersects line AB at point M and that T is on line segment PM, $PT = \lambda PQ$, then

$$\triangle TAB = (1-\lambda) \triangle PAB - \lambda \triangle QAB$$

Proof: Applying the Co-Side Theorem to Fig. 4.8

we have
$$\frac{\triangle PAB}{PM} = \frac{\triangle TAB}{TM} = \frac{\triangle QAB}{QM}$$

thus
$$\frac{\triangle PAB - \triangle TAB}{PM - TM} = \frac{\triangle TAB + \triangle TAB}{TM + QM}$$

namely
$$\frac{\triangle PAB - \triangle TAB}{\triangle TAB + \triangle TAB} = \frac{PM - TM}{TM + QM} = \frac{PT}{TQ} = \frac{\lambda}{1 - \lambda}$$

solving for $\triangle TAB$
$$\triangle TAB = (1 - \lambda) \triangle PAB - \lambda \triangle QAB$$

ADDITIONAL PROBLEMS

[P4.1] In Fig. 4.1, let $ABCD$ be trapezoid and $AB \parallel CD$.

Given: $CD = \frac{2}{3} AB$ (CD is the upper base), the area of $ABCD$ is S. Determine the area of the nine pieces cut by the Chinese character '井'.

[P4.2] Replacing the trisections points with points dividing equally into four parts and points dividing equally into six parts in Fig. 4.1, what results can you get?

[P4.3] Think: in Fig. 4.1, given the area of $\triangle ABD$, $\triangle ABC$ and $\triangle BCD$, can you determine the area of each of the 9 pieces divided by the character '井'?

Chapter 5

Killing Three Birds with One Stone

Let's talk about an interesting example. Although it seems hard at first glance, there exists a very smart and simple solution. What's amazing is that this solution can be found from a sequence of simple steps.

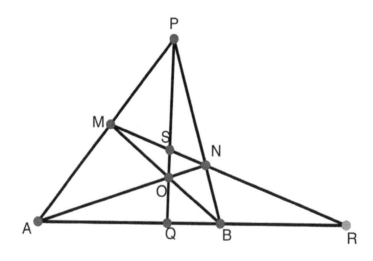

FIGURE 5.1

Chapter three (Ex. 3.5) introduced a mathematics competition problem suggested by Buqing Su in his letter to Luogeng Hua. In this problem, there were two parallel lines: *AB* and *MN*. If *AB* and *MN* are <u>not</u> parallel, with an intersection point *R* (As shown in Fig. 5.1), then point *Q* is obviously not the midpoint. In this case, is there anything to mention about the position of *Q* on *AB*?

In fact, Luogeng Hua has talked about this problem in the preface of the book—Solutions to Middle School Mathematics Competitions, China,1978. He pointed out that although $AQ = QB$ no longer holds in the situation shown in Fig. 5.1, the following equation is true:

$$\frac{AQ}{BQ} = \frac{AR}{BR} \tag{5.1}$$

Besides, as point R moves away from point B, $\frac{AR}{BR}$ approaches 1. When $MN \parallel AB$, we may imagine that R reaches infinity, that is $\frac{AR}{BR} = 1$.

So Fig. 5.1 is the general case of Fig.3.5.

Luogeng Hua pointed out that the problem suggested by Buqing Su contained the basic principle of affine geometry. And the equation (5.1), the general case in Fig. 5.1 holds, which contains the basic principle of projective geometry. In the *Solutions* preface, Luogeng Hua has given a proof that can be understood by middle school students. The proof is one page long and uses the knowledge of trigonometric functions. We don't repeat the proof here. Interested readers can refer to the preface of the book, *Solutions to Middle School Mathematics Competitions*, China,1978.

By applying the Co-Side Theorem, we can get the general result with a simpler approach.

[Ex5.1] As shown in Fig. 5.1, try proving the following equations:

$$\frac{AQ}{BQ} = \frac{AR}{BR} \tag{5.1}$$

$$\frac{PS}{OS} = \frac{PQ}{OQ} \tag{5.2}$$

$$\frac{MS}{NS} = \frac{MR}{NR} \qquad\qquad (5.3)$$

Proof: First, prove equation (5.1). Obviously, proving

$\dfrac{AQ}{BQ} \cdot \dfrac{BR}{AR} = 1$ is enough.

According to the Co-Side Theorem, we have:

$$\begin{aligned}
\frac{AQ}{BQ} \cdot \frac{BR}{AR} &= \frac{\triangle AOP}{\triangle BOP} \cdot \frac{\triangle BMN}{\triangle AMN} \\
&= \frac{\triangle AOP}{\triangle ABP} \cdot \frac{\triangle ABP}{\triangle BOP} \cdot \frac{\triangle BMN}{\triangle OMN} \cdot \frac{\triangle OMN}{\triangle AMN} \\
&= \frac{MO}{MB} \cdot \frac{NA}{NO} \cdot \frac{MB}{MO} \cdot \frac{NO}{NA} = 1
\end{aligned}$$

Thus, equation (5.1) has been proved.

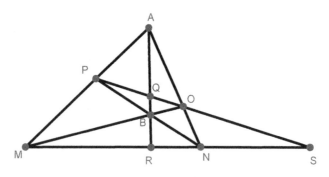

FIGURE 5.2

Maybe you think only one third of the work has been done. In fact, the equation (5.2) and (5.3) have already been proved. The deduction above can be applied to prove (5.2) and (5.3) simply by exchanging the letters in Fig. 5.1.

Fig. 5.2 and Fig. 5.1 are the same except for the point labels. The proof of (5.1) also applies to Fig. 5.2. So it is proved in Fig. 5.2 that

$\dfrac{AQ}{BQ} = \dfrac{AR}{BR}$ which means that $\dfrac{PS}{OS} = \dfrac{PQ}{OQ}$ is also true for Fig. 5.1.

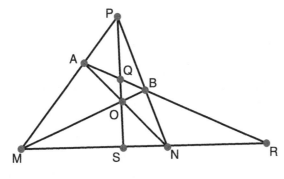

FIGURE 5.3

Rearrange the point labels again as shown in Fig. 5.3. The proof of (5.1) also applies to Fig. 5.3!

So, in Fig. 5.3, we also have:
$$\frac{AQ}{BQ} = \frac{AR}{BR}$$

Comparing the letters in Fig. 5.3 with Fig. 5.1, you will find that $\dfrac{MS}{NS} = \dfrac{MR}{NR}$

is also true in Fig. 5.1.

Exchanging these letters and explaining them repeatedly may be not as good as presenting the proofs of (5.2) and (5.3) directly. But this method is not only interesting but also inspires us to further explore the reason why it works.

It's not a coincidence. The truth will come out as long as readers are told how to draw the figure in the problem. Ex. 5.1 should not simply state: "As shown in Fig. 5.1". Instead, the statement should be replaced by the following:

"Take any four points *M*, *N*, *A* and *B* in a plane. Line *AB* and *MN* intersect at point *R*. Line *MA* and *NB* intersect at point *P*. Line *NA* and *MB* intersect at point *O*. Lines *PO* and *AB* intersect at point *Q*. Line *PO* and *MN* intersect at *S*.

Prove:
$$\frac{AQ}{BQ} = \frac{AR}{BR} \quad \text{........}$$

The problem doesn't rely on the figure when described as above. The problem starts with "Take any four points *M, N, A* and *B* in a plane", which means that when these four points are determined, the figure is fixed. If the relative position of four points is different, the position of line segments in the conclusion is also different in the figure. The difference between Fig. 5.1, Fig. 5.2 and Fig. 5.3 lies in the fact that although the relative position of point *M, N, A* and *B* changes, their relation with other points stays the same. *R* is still the intersection point of *AB* and *MN*. *P* is still the intersection point of *MA* and *NB*. *O* is still the intersection point of *NA* and *MB*. *Q* is still the intersection point of *PO* and *AB*. *S* is still the intersection point of *PO* and *MN*. Throughout the proof, we applied the Co-Side Theorem. The Co-Side Theorem does not rely on a specific figure. It's only concerned with the condition that a certain point is the intersection point of two lines. Therefore, the proof of one figure is also available for others since the relation of intersection points are the same in different figures.

This is why our proof can kill three birds with one stone.

In problems like this, we may get dazzled if we only look at the figure. Given that the mathematician's proof is very long, how can we find this simple and multi-functional method?

Given below is the thought process. It is more important than the problem itself. Grasp it, and you can solve quite a few problems that seem to have few clues at first glance.

Before starting your problem, sequence the points in the figure by their stated order:

The first batch of points are *A, B, M* and *N*. They are free points without restraints. The only condition is any three of them cannot be on a line, otherwise the figure cannot be drawn.

The second batch of points is *P, R* and *O*. They are determined by the first batch of points:

> *P* is the intersection point of *AM* and *BN*;
>
> *R* is the intersection point of *AB* and *MN*;
>
> *O* is the intersection point of *AN* and *BM*.

The third batch of points is Q and S. They are determined by the previous two batches of points:

> S is the intersection point of PO and MN;

> Q is the intersection point of PO and AB.

The first batch of points is called the **free points**, and the latter two batches are called **constraint points**.

The steps of a proof have a strong relationship with their sequence.

The conclusion is $\dfrac{AQ}{BQ} = \dfrac{AR}{BR}$, which means we need to prove $\dfrac{AQ}{BQ} \cdot \dfrac{BR}{AR} = 1$.

So, what we care about is the formula $\dfrac{AQ}{BQ} \cdot \dfrac{BR}{AR}$.

How to deal with it?

"**Elimination**" is used to solve linear equations with multiple variables. When the unknowns are eliminated one by one, the problem will be solved. This method can also be applied to geometry and let's just call it **The Eliminating Points Method**.

The Eliminating Points Method eliminates the constraint points in our formula. When all constraint points are eliminated, the conclusion appears like a rock emerging as the water level drops.

Eliminating constraint points must comply with this rule: last generated, first eliminated. In the

formula $\dfrac{AQ}{BQ} \cdot \dfrac{BR}{AR}$, Q is the last generated, so it will be eliminated first.

How do we eliminate point Q? We must find out where Q comes from. This is called "whoever started the trouble should end it". Once it's identified, we know Q is the intersection point of PO and AB.

According to the condition and applying the Co-Side Theorem, we have:

$$\frac{AQ}{BQ} = \frac{\triangle APO}{\triangle BPO}.$$

So the formula $\dfrac{AQ}{BQ} \cdot \dfrac{BR}{AR}$ transforms to the following:

$$\frac{AQ}{BQ} \cdot \frac{BR}{AR} = \frac{\triangle APO}{\triangle BPO} \cdot \frac{BR}{AR} \qquad (5.4)$$

Next, we should eliminate P, O and R from the right-hand side of (5.4). It's easy to eliminate R since AB intersects MN at point R. So, we have

$$\frac{BR}{AR} = \frac{\triangle BMN}{\triangle AMN} \qquad (5.5)$$

But it's difficult to eliminate P and O from $\triangle APO$ and $\triangle BPO$. Now, there is both an awkward method and a smart method.

The smart one is to make a bridge by using a triangle which has common sides with $\triangle APO$ and $\triangle BPO$. For example, using $\triangle ABO$ as a transition, we have:

$$\frac{\triangle APO}{\triangle BPO} = \frac{\triangle APO}{\triangle ABO} \cdot \frac{\triangle ABO}{\triangle BPO} = \frac{PN}{BN} \cdot \frac{AM}{PM}$$

$$= \frac{\triangle PMN}{\triangle BMN} \cdot \frac{\triangle AMN}{\triangle PMN} = \frac{\triangle AMN}{\triangle BMN} \qquad (5.6)$$

Substituting (5.6) and (5.5) into (5.4), then we have solved the problem.

The awkward one is to eliminate P and Q one by one and deal with these two triangles respectively.

$$\begin{cases} \triangle APO = \dfrac{AP}{AM} \cdot \triangle AMO = \dfrac{AP}{AM} \cdot \dfrac{AO}{AN} \cdot \triangle AMN \\ \dfrac{AP}{AM} = \dfrac{\triangle ABN}{\triangle ABN - \triangle BMN} \\ \dfrac{AO}{AN} = \dfrac{\triangle ABM}{\triangle BMN + \triangle ABM} \end{cases} \qquad (5.7)$$

$$\begin{cases} \triangle BPO = \dfrac{BP}{BN} \cdot \triangle BNO = \dfrac{BP}{BN} \cdot \dfrac{BO}{BN} \cdot \triangle BMN \\[4mm] \dfrac{BP}{BN} = \dfrac{\triangle ABM}{\triangle ABM - \triangle AMN} \\[4mm] \dfrac{BO}{BM} = \dfrac{\triangle ABN}{\triangle AMN + \triangle ABN} \end{cases}$$ (5.8)

Substitute (5.5), (5.7) and (5.8) into (5.4), and notice that

$$\triangle AMN + \triangle ABN = \triangle BMN + \triangle ABM$$

$$\triangle ABN - \triangle BMN = \triangle ABM - \triangle AMN$$

Similarly, the problem is solved.

By the way, in (5.7) and (5.8), here's how $\dfrac{AP}{AM}, \dfrac{AO}{AN}, \dfrac{BP}{BN}$ and $\dfrac{BO}{BM}$ are turned into ratios of areas. Look at them upside down:

$$\begin{aligned} \frac{AM}{AP} &= \frac{AP - MP}{AP} \\ &= 1 - \frac{MP}{AP} = 1 - \frac{\triangle BMN}{\triangle ABN} \\ &= \frac{\triangle ABN - \triangle BMN}{\triangle ABN} \end{aligned}$$

$$\begin{aligned} \frac{AN}{AO} &= \frac{ON + AO}{AO} \\ &= \frac{ON}{AO} + 1 = \frac{\triangle BMN}{\triangle ABM} + 1 \\ &= \frac{\triangle BMN + \triangle ABM}{\triangle ABM} \end{aligned}$$

and they are determined. The other two formulas are tackled likewise.

By comparison, the method using transition is much better than the other. The transition technique is used repeatedly with the Co-Side Theorem (1.6) in the proofs of Ex. 2.1, Ex. 2.2, Ex. 2.6, Ex. 3.5. This technique deserves careful consideration. However, in case there is no other smart method, or you cannot come up with one, the awkward approach may also be the key to the solution. The advantage of the awkward approach is that you can solve problems step by step without cudgeling your brains.

ADDITIONAL PROBLEMS

[P5.1] Given the conditions in Ex. 5.1: a new line BS which intersects ON at point X and PM at Y. Then, are there any new proportional line segments? Please find out and prove them.

[P5.2] In Ex. 5.1, if $\triangle PON$ is a transition triangle, try finishing the following deduction:

$$\frac{AQ}{BQ} = \frac{\triangle AOP}{\triangle BOP} = \frac{\triangle AOP}{\triangle PON} \cdot \frac{\triangle PON}{\triangle BOP}$$

$$= \frac{(\ )}{(\ )} \cdot \frac{(\ )}{(\ )} = \frac{\triangle ABM}{(\ )} \cdot \frac{(\ )}{(\ )}$$

$$= \frac{(\ )}{(\ )} = \frac{AR}{BR}$$

[P5.3] Please find the solutions to Ex. 2.4, Ex. 2.7 and Ex. 3.5 by the Eliminating Points Method.

Chapter 6

Proving the Pappus Theorem and the Gaussian Line Theorem by the Eliminating Points Method

We show how the Eliminating Points Method works with two examples.

[Ex6.1] Given: points *A*, *B* and *C* are collinear and so are points *X*, *Y* and *Z*. Lines *BX* and *AY*, *BZ* and *CY*, *AZ* and *CX* intersect at *P*, *Q*, *R*, respectively (as shown in Fig. 6.1).

Prove: *P*, *Q* and *R* are collinear.

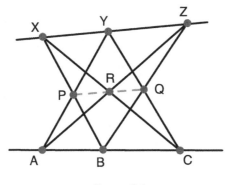

Figure 6.1

This example is the famous **Pappus Theorem**. This is a beautiful theorem! Yet, we may not know how to prove it at first glance. Now let's apply the Eliminating Points Method and see how it works.

The problem is how to prove three points lie on a line. The question is not concrete enough. So, let's specify it.

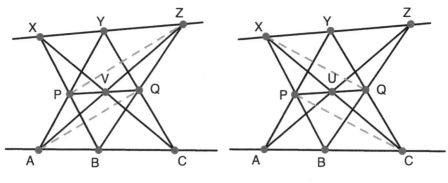

FIGURE 6.2

To prove that points *P*, *Q* and *R* are on a line means to prove that point *R* is on line *PQ*. Since *R* is the intersection point of *CX* and *AZ*, we need to prove *CX*, *AZ* and *PQ* have a common intersection point in order to prove the original proposition. To make it clear, suppose that *CX* intersects *PQ* at *U* and *AZ* intersects *PQ* at *V*. As shown in Fig. 6.2, it is

sufficient as long as we prove $\dfrac{PU}{QU} = \dfrac{PV}{QV}$, that is,

Prove: $\dfrac{PU}{QU} \cdot \dfrac{QV}{PV} = 1$

Let's order these points as follows:

The first batch of points: *A*, *B* and *C* are collinear and so are *X*, *Y* and *Z*.

The second batch of points: *P* and *Q*.

> *P* — the intersection point of *AY* and *BX*;

> *Q* — the intersection point of *BZ* and *CY*;

> The third batch of points: *U* and *V*.

> *U* — the intersection point of *PQ* and *XC*;

> *V* — the intersection point of *PQ* and *AZ*;

According to Eliminating Points Method, we should eliminate U and V from (6.1)'s left-hand side. Based on the derivations of U and V, we can obtain the following equations by the Co-Side Theorem:

$$
\begin{cases}
\dfrac{PU}{QU} = \dfrac{\triangle PXC}{\triangle QXC} \\[3mm]
\dfrac{QV}{PV} = \dfrac{\triangle QAZ}{\triangle PAZ}
\end{cases}
\tag{6.2}
$$

Moreover, let's eliminate the second batch of points, that is to eliminate point P and Q from (6.2)'s right-hand side:

$$
\begin{cases}
\triangle PXC = \dfrac{\triangle PXC}{\triangle BXC} \cdot \triangle BXC = \dfrac{PX}{BX} \cdot \triangle BXC = \dfrac{PX \cdot \triangle BXC}{BP + PX} \\[3mm]
\qquad\quad = \dfrac{\triangle AXY \cdot \triangle BXC}{\triangle ABY + \triangle AXY} = \dfrac{\triangle AXY \cdot \triangle BXC}{S_{ABXY}} \\[3mm]
\triangle QXC = \dfrac{\triangle QXC}{\triangle XYC} \cdot \triangle XYC = \dfrac{QC}{YC} \cdot \triangle XYC = \dfrac{\triangle BCZ \cdot \triangle XYC}{S_{BCZY}}
\end{cases}
\tag{6.3}
$$

$$
\begin{cases}
\triangle QAZ = \dfrac{\triangle QAZ}{\triangle ABZ} \cdot \triangle ABZ = \dfrac{QZ}{BZ} \cdot \triangle ABZ = \dfrac{\triangle YZC \cdot \triangle ABZ}{S_{BCZY}} \\[3mm]
\triangle PAZ = \dfrac{\triangle PAZ}{\triangle AYZ} \cdot \triangle AYZ = \dfrac{AP}{AY} \cdot \triangle AYZ = \dfrac{\triangle ABX \cdot \triangle AYZ}{S_{ABYX}}
\end{cases}
\tag{6.4}
$$

Now the first batch of points remains. Let's take a look at the results. Substituting (6.4) and (6.3) into (6.2) first and then into (6.1), we have:

$$
\frac{PU}{QU} \cdot \frac{QV}{PV} = \frac{\triangle PXC}{\triangle QXC} \cdot \frac{\triangle QAZ}{\triangle PAZ}
$$

$$
= \frac{\triangle AXY \cdot \triangle BXC}{S_{ABYX}} \cdot \frac{S_{BCZY}}{\triangle BCZ \cdot \triangle XYC} \cdot \frac{\triangle YZC \cdot \triangle ABZ}{S_{BCZY}} \cdot \frac{S_{ABYX}}{\triangle ABX \cdot \triangle AYZ}
$$

$$= \frac{\triangle AXY}{\triangle AYZ} \cdot \frac{\triangle BXC}{\triangle ABX} \cdot \frac{\triangle YZC}{\triangle XYC} \cdot \frac{\triangle ABZ}{\triangle BCZ}$$

$$= \frac{XY}{YZ} \cdot \frac{BC}{AB} \cdot \frac{YZ}{XY} \cdot \frac{AB}{BC} = 1$$

This is the proof of the Pappus Theorem.

[Ex6.2] In quadrilateral *ABCD*, opposite sides *DA* and *CB* are extended to intersect at *K*; *AB* and *DC* are extended to intersect at *L*. The midpoints of two diagonals *AC* and *BD* are *N* and *M* respectively. Line *MN* intersects *KL* at *P*.

Prove: $KP = PL$

Another version of this question is to prove the point *M*, *N* and the midpoint *P* of *LK* are collinear. This line is called the **Gaussian line** of quadrilateral *ABCD*.

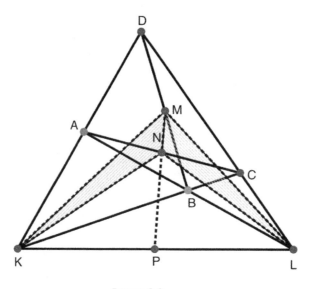

FIGURE 6.3

As shown in Fig. 6.3, the points are grouped as follows:

The first batch of points: *A, B, C, D* are free points.

In the second batch of points, *M, N, K, L*:

M is the midpoint of *BD*, and *K* is the intersection point of *DA* and *BC*.

N is the midpoint of AC, and L is the intersection point of AB and DC.

In the third batch: point P is the intersection point of MN and KL.

When applying the Eliminating Points Method, first notice that what we need to prove is

$$\frac{KP}{PL} = 1$$

So first of all you should eliminate point P from $\dfrac{KP}{PL}$ since point P is the intersection

point of MN and KL, that is

$$\frac{KP}{PL} = \frac{\triangle KMN}{\triangle LMN} \tag{6.5}$$

Next, we will gradually eliminate point K, L, M and N from (6.5)'s right-hand side. By the **_Supplement of Point of Division Formula_** introduced previously, we have: (we make use of $AN = NC = \frac{1}{2}AC$)

$$\begin{cases} \triangle KMN = \dfrac{1}{2}(\triangle KMC - \triangle KMA) \\ \triangle LMN = \dfrac{1}{2}(\triangle LMA - \triangle LMC) \end{cases} \tag{6.6}$$

Then, with the condition that $BM = MD = \frac{1}{2}BD$, we have:

$$\begin{cases} \triangle KMC = \dfrac{1}{2}\triangle KDC, \quad \triangle KMA = \dfrac{1}{2}\triangle KBA \\ \triangle LMA = \dfrac{1}{2}\triangle LDA, \quad \triangle LMC = \dfrac{1}{2}\triangle LBC \end{cases} \tag{6.7}$$

Substituting (6.7) into (6.6), we have:

47

$$\begin{cases} \triangle KMN = \dfrac{1}{4}(\triangle KDC - \triangle KBA) = \dfrac{1}{4}S_{ABCD} \\ \triangle LMN = \dfrac{1}{4}(\triangle LDA - \triangle LBC) = \dfrac{1}{4}S_{ABCD} \end{cases} \tag{6.8}$$

Then, substituting it into (6.5), we obtain the desired conclusion.

Here, to eliminate point K and L, we use a coincident situation that the difference of area of two triangles is the area of quadrilateral. Generally, we can use an awkward method:

$$\triangle KDC = \frac{KD}{AD} \cdot \triangle ADC = \frac{KD \cdot \triangle ADC}{KD - KA} = \frac{\triangle BDC \cdot \triangle ADC}{\triangle BDC - \triangle ABC}$$

$$\triangle KBA = \frac{KA}{AD} \cdot \triangle ABD = \frac{KA \cdot \triangle ABD}{KD - KA} = \frac{\triangle ABC \cdot \triangle ABD}{\triangle BDC - \triangle ABC} \tag{6.9}$$

$$\triangle LDA = \frac{LD}{CD} \cdot \triangle ADC = \frac{LD \cdot \triangle ADC}{LD - LC} = \frac{\triangle ABC \cdot \triangle ADC}{\triangle ABD - \triangle ABC}$$

$$\triangle LBC = \frac{LC}{DC} \cdot \triangle BDC = \frac{LC \cdot \triangle BDC}{LD - LC} = \frac{\triangle ABC \cdot \triangle BDC}{\triangle ABD - \triangle ABC}$$

Then, substituting $\triangle BDC = \triangle ABC + \triangle ADC - \triangle ABD$ into (6.9), we have:

$$\begin{aligned} \triangle KDC - \triangle KBA &= \frac{\triangle BDC \cdot \triangle ADC - \triangle ABC \cdot \triangle ABD}{\triangle BDC - \triangle ABC} \\ &= \frac{(\triangle ADC - \triangle ABD) \cdot (\triangle ABC + \triangle ADC)}{\triangle ADC - \triangle ABD} \\ &= S_{ABCD} \end{aligned}$$

Similarly, we have $\triangle LDA - \triangle LBC = S_{ABCD}$. The problem is solved.

Comparing the smart approach and the awkward one, we perceive that carefully observing the figures can be very useful for solving geometric problems. Something that can be directly observed from the figure can instead be complicated in calculation and even require factorization when one works it out algebraically.

Chapters 5 and 6 illustrate the utility of the Eliminating Points Method. The key steps to using the Eliminating Points Method are:

1. Order points in the problem by the order of their occurrence. First generated, first sequenced.
2. Transform the problem into one concerned with simplification of a certain equation. Some geometric quantities will be involved in the equation.
3. Eliminate points generated by constraint conditions step by step in the simplified equation. Last generated, first eliminated.
4. When eliminating a point, apply geometric conditions applied to the point. Make sure to find out the shortcuts implied by the figure.

After you have mastered this method, you will be confident when confronting geometry problems.

ADDITIONAL PROBLEMS

 [P6.1] As shown in Figure 6.4, *D* is any point in the interior of Δ*ABE*. Point *C* is any point on *AB*. Line *AE* intersects *CD* at point *J* and *AE* intersects *BD* at point *I*. Line *AD* intersects *CE* at point *H* and *AD* intersects *BE* at point *G*. Now prove by the Eliminating Points Method: *IH*, *JG* and *AB* intersect at a point.

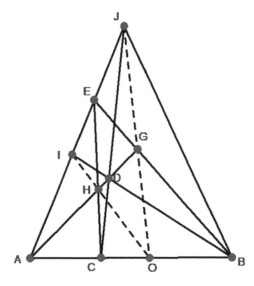

FIGURE 6.4

CHAPTER 7

CO-ANGLE TRIANGLES AND THE CO-ANGLE THEOREM

To study geometric figures, we should emphasize those basic figures which are frequently encountered. The co-side triangle is one such figure. When investigating it, we found that the Co-Side Theorem proved very fruitful.

Yet, the conditions and conclusions of the Co-Side Theorem do not mention any angles, so it cannot help us solve any geometric problems involving angles. To solve these problems, we need to find some new tools.

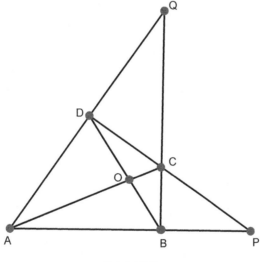

FIGURE 7.1

Consider a geometric figure that consists of any four points, as shown in Fig. 7.1. We know that there are many co-side triangles in the figure. Now, let's focus on another kind of relationship between triangles. For example, $\triangle AOD$ and $\triangle BOC$ are not co-side triangles, but they have another relation: $\angle AOD = \angle BOC$. On the other hand, consider

$\triangle BOC$ and $\triangle AQC$, in this case the triangles have supplementary angles $\angle BCO$ and $\angle ACQ$.

We call this kind of triangles **Co-Angle Triangles**. Precisely, if two angles $\angle ABC$ and $\angle A'B'C'$ are equal or supplementary, we may say $\triangle ABC$ and $\triangle A'B'C'$ are a pair of **Co-Angle Triangles**.

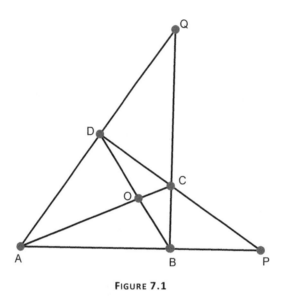

FIGURE 7.1

In Fig. 7.1, there are many pairs of co-angle triangles: $\triangle AQB$ and $\triangle DQC$, $\triangle AOB$ and $\triangle DOC$, $\triangle ABO$ and $\triangle AOD$, $\triangle BCD$ and $\triangle DCQ$, $\triangle AOB$ and $\triangle BPD$, etc. The number of co-angle triangles is not necessarily less than the number of co-side triangles. For $\triangle AOD$, there are five triangles having common sides with it: $\triangle ADP$, $\triangle ADC$, $\triangle ADB$, $\triangle AOB$ and $\triangle DOC$. Whereas, there are seven triangles having common angles with it: $\triangle ADC$, $\triangle QAC$, $\triangle ABD$, $\triangle BOC$, $\triangle AOB$, $\triangle DOC$ and $\triangle QBD$. It is clear that studying co-angle triangles is of great importance.

A pair of co-side triangles, of course, always attach to each other by their common side. But co-angle triangles can be apart. Congruent triangles and similar triangles, which are familiar to us, are both special cases of co-angle triangles. So, co-angle triangles are more varied and useful. As co-side triangles have the Co-Side Theorem, co-angle triangles have the Co-Angle Theorem

Co-Angle Theorem: If $\angle ABC$ and $\angle A'B'C'$ are equal or supplementary, then we have:

$$\frac{\triangle \text{ABC}}{\triangle \, A'B'C'} = \frac{AB \cdot BC}{A'B' \cdot B'C'}$$

Proof: As shown in Fig. 7.2, merge the two triangles so that the two lines of $\angle B$ and $\angle B'$ coincide, where (A) shows the situation that the two angles are equal and (2) shows the situation that the two angles are supplementary. Under these two situations, we have:

$$\frac{\triangle \, ABC}{\triangle \, A'B'C'} = \frac{\triangle \, ABC}{\triangle \, A'BC} \cdot \frac{\triangle \, A'BC}{\triangle \, A'B'C'}$$

$$= \frac{AB \cdot BC}{A'B' \cdot B'C'}$$

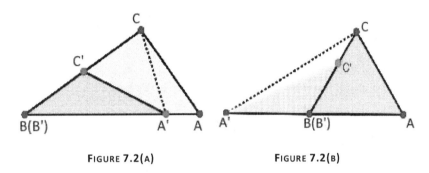

FIGURE 7.2(A)　　　　　**FIGURE 7.2(B)**

Although the acquisition of the theorem takes a little effort, it proves very useful. Let's illustrate with some simple examples.

[Ex7.1]　In $\triangle ABC$, $\angle B = \angle C$.

Prove: $AB = AC$.

Proof: Take $\triangle ABC$ and $\triangle ACB$ as two triangles. By the Co-Angle Theorem, we have:

$$1 = \frac{\triangle \, ABC}{\triangle \, ACB} = \frac{AB \cdot BC}{AC \cdot BC}$$

$$\therefore \quad \frac{AB}{AC} = 1$$

Although this example is simple, it shows the advantage of this method. We needn't construct congruent triangles. Even without drawing the figure the problem has been solved.

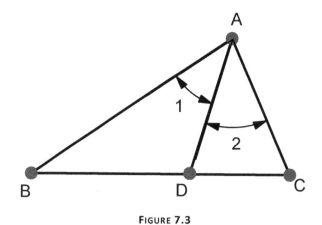

FIGURE 7.3

[Ex7.2] As shown in Fig. 7.3, given that AD is an angle bisector of $\triangle ABC$.

prove:

$$\frac{BD}{CD} = \frac{BA}{CA}$$

.

Proof: Since $\angle BAD \;=\; \angle CAD$, we have:

$$\frac{BD}{CD} = \frac{\angle BAD}{\angle CAD} = \frac{BA \cdot AD}{CA \cdot AD} = \frac{BA}{CA}$$

[Ex7.3] In $\triangle ABC$ and $\triangle A'B'C'$, $\angle A = \angle A'$, $\angle B = \angle B'$.

Prove:

$$\frac{AB}{A'B'} = \frac{BC}{B'C'} = \frac{CA}{C'A'}$$

Proof: From the given, it's obvious that $\angle C = \angle C'$. Applying the Co-Angle Theorem three times to $\angle ABC$ and $\triangle A'B'C'$, we have:

$$\frac{\triangle ABC}{\triangle A'B'C'} = \frac{AB \cdot BC}{A'B' \cdot B'C'} = \frac{BC \cdot CA}{B'C' \cdot C'A'} = \frac{CA \cdot AB}{C'A' \cdot A'B'}$$

Since $\qquad \dfrac{AB \cdot BC}{A'B' \cdot B'C'} = \dfrac{BC \cdot CA}{B'C' \cdot C'A'}$

we have
$$\frac{AB}{A'B'} = \frac{CA}{C'A'}$$

Since
$$\frac{BC \cdot CA}{B'C' \cdot C'A'} = \frac{CA \cdot AB}{C'A' \cdot A'B'}$$

we have
$$\frac{BC}{B'C'} = \frac{AB}{A'B'}$$

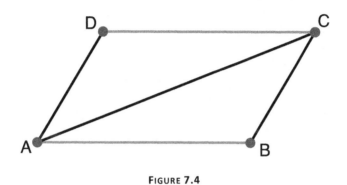

FIGURE 7.4

[Ex7.4] In parallelogram $ABCD$, **prove:** $AB = CD$.

Proof: Since $\triangle ABC = \triangle ADC$ and $\angle BAC = \angle DCA$, according to the Co-Angle Theorem, we have:

$$1 = \frac{\triangle BAC}{\triangle DCA} = \frac{AB \cdot AC}{CD \cdot AC} = \frac{AB}{CD}$$

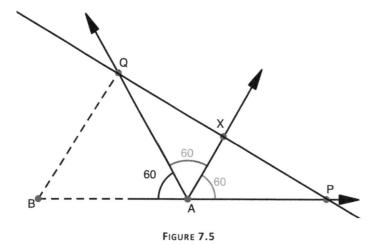

FIGURE 7.5

[Ex7.5] As shown in Fig. 7.5, three rays through point A construct two angles that are 60°. A line intersects them at points P, X and Q.

Prove: $$\frac{1}{AX} = \frac{1}{AP} + \frac{1}{AQ}$$

Proof: Take any point B on the extension line of PA.

$\triangle PAQ = \triangle PAX + \triangle QAX$ holds certainly.

Dividing the two sides of the equation above by $\triangle BAQ$, we have

$$\frac{\triangle PAQ}{\triangle BAQ} = \frac{\triangle QAX}{\triangle BAQ} + \frac{\triangle PAX}{\triangle BAQ}$$

According to the Co-Angle Theorem, we have

$$\frac{PA \cdot AQ}{AB \cdot AQ} = \frac{AX \cdot AQ}{AB \cdot AQ} + \frac{AP \cdot AX}{AB \cdot AQ}$$

Multiply the two sides of the equation above by AB, and divide them by $AX \cdot AP$ Then, the result is:

$$\frac{1}{AX} = \frac{1}{AP} + \frac{1}{AQ}$$

Used in the proof of Ex7.5 is this Area Equation: $\triangle PAQ = PAX + QAX$. This is one of the typical techniques of solving geometric problems. Along with Co-Side Theorem and

Co-Angle Theorem, it can solve a lot of problems. Later on, we will illustrate this with more examples including some rather complicated problems and even some sourced from international mathematics competitions.

By the way, Ex7.5 provides a simple graphical approach to solve an equation of the form:

$$\frac{1}{x} = \frac{1}{a} + \frac{1}{b}$$

Given: $\frac{1}{x} = \frac{1}{5} + \frac{1}{4}$. Determine *x*. It is troublesome to calculate it by hand. Now: Take *AQ* = 4cm, *AP* = 5cm, and connect *P* and *Q*. Point *X* is their intersection point. Then, by measuring *AX*, we can get the approximate of *x*. This kind of calculation is very useful in Physics because the *Parallel Resistance Formula* is

$$\frac{1}{R} = \frac{1}{R_1} + \frac{1}{R_2}$$

and the *Series Capacitance Formula* is

$$\frac{1}{C} = \frac{1}{C_1} + \frac{1}{C_2}$$

In Optics, the *Focal Length Formula of a Lens* is:

$$\frac{1}{f} = \frac{1}{f_1} + \frac{1}{f_2}$$

All of these can use the same graphical solution.

Now, the following examples are a bit more complicated.

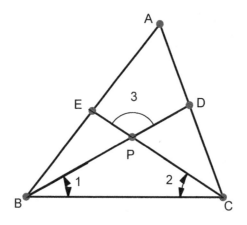

FIGURE 7.6

[Ex7.6] In $\triangle ABC$, point D and E are on side AC and AB respectively such that

$$\angle DBC = \angle ECB = \frac{\angle A}{2}.$$

Prove: $BE = CD.$

Proof: Denote the intersection point of BD and CE as P. As shown in Fig. 7.6, $\angle A$ and $\angle 3$ are supplementary since $\angle 1 + \angle 2 = \angle A$. Therefore, $\angle BEC$ and $\angle CDB$ are supplementary. According to Co-Angle Theorem, we have

$$\frac{BE \cdot CE}{CE \cdot CD} = \frac{\triangle BEC}{\triangle BDC} = \frac{BC \cdot CE}{BD \cdot BC}$$

by reduction,

$$\frac{BE}{CD} = 1$$

[Ex7.7] (Liao Ning Middle School Mathematics Competition, China, 1978) In $\triangle ABC$, point M is the midpoint of side BC. Take point E and F on AB and AC respectively. Connect E, F and connect A, M such that EF intersects AM at point N.

Prove:
$$\frac{AM}{AN} = \frac{1}{2}\left(\frac{AB}{AE} + \frac{AC}{AF}\right)$$

Proof: As $MB = MC$, we have:

$$\triangle ABC = 2\,\triangle ABM = 2\,\triangle ACM \qquad (7.1)$$

Divide both sides of the following equation

$$\triangle AEF = \triangle AEN + \triangle AFN \qquad (7.2)$$

by $\triangle ABC$. With (7.1), we have:

$$\frac{\triangle AEF}{\triangle ABC} = \frac{\triangle AEN}{2 \triangle ABM} + \frac{\triangle AFN}{2 \triangle ACM} \qquad (7.3)$$

Apply Co-Angle Theorem to (3), we have

$$\frac{AE \cdot AF}{AB \cdot AC} = \frac{1}{2}\left(\frac{AE \cdot AN}{AB \cdot AM} + \frac{AF \cdot AN}{AC \cdot AM}\right) = \frac{1}{2}\left(\frac{AE}{AB} + \frac{AF}{AC}\right) \cdot \frac{AN}{AM} \qquad (7.4)$$

Resolving (7.4), we get

$$\frac{AM}{AN} = \frac{1}{2}\left(\frac{AB}{AE} + \frac{AC}{AF}\right)$$

To solve for $\dfrac{AM}{AN}$ from (7.4), multiply both sides of (7.4) by the following expression.

$$\frac{AB \cdot AC \cdot AM}{AE \cdot AF \cdot AN}$$

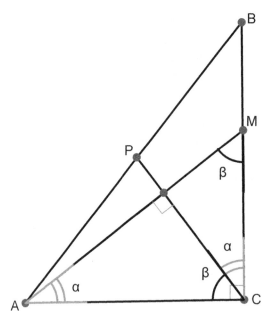

[Ex7.8] Given: $\triangle ABC$ is an isosceles right triangle and $\angle C = 90°$. Take point M on BC such that $CM = 2MB$. Draw a line perpendicular to MA from C, intersecting AB at P (as shown in Fig. 7.7).

Determine:

$$\frac{AP}{PB}$$

.

Solution: Notice that $\angle ACP = \angle AMC$ and $\angle BCP = \angle MAC$. The following deduction can be done with Co-Angle Theorem:

$$\frac{AP}{PB} = \frac{\triangle ACP}{\triangle BCP} = \frac{\triangle ACP}{\triangle AMC} \cdot \frac{\triangle AMC}{\triangle BCP}$$

$$= \frac{AC \cdot CP}{AM \cdot MC} \cdot \frac{AM \cdot AC}{BC \cdot CP}$$

$$= \frac{AC \cdot AC}{MC \cdot BC} = \frac{3}{2}$$

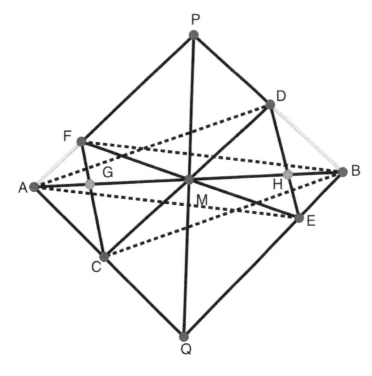

FIGURE 7.8

[Ex7.9] The *Quadrilateral Butterfly Theorem*

Given: in quadrilateral *AQBP*, the diagonal line *PQ* crosses the midpoint M of the other diagonal AB. Draw two lines through *M: one* intersects *AQ* and *BP* at *C* and *D* respectively; *the other* intersects *BQ* and *AP* at *E* and *F* respectively. Connect *CF* and *DE*, which intersect *AB* at *G* and *H* respectively (as shown in Fig. 7.8).

Prove: *MG = MH.*

Proof: Since *AM = BM*, it is sufficient to prove

$$\frac{MG}{AG} = \frac{MH}{BH},$$

which can be transformed to prove the following equation:

$$\frac{MG}{AG} = \frac{BH}{MH} = 1$$

whereas,

$$\frac{MG}{AG} \cdot \frac{BH}{MH} = \frac{\triangle MCF}{\triangle ACF} \cdot \frac{\triangle BDE}{\triangle MDE}$$

$$= \frac{\triangle MCF}{\triangle MDE} \cdot \frac{\triangle BDE}{\triangle BPQ} \cdot \frac{\triangle BPQ}{\triangle APQ} \cdot \frac{\triangle APQ}{\triangle ACF}$$

$$= \frac{MC \cdot MF}{MD \cdot ME} \cdot \frac{BD \cdot BE}{BP \cdot BQ} \cdot \frac{BM \cdot BQ}{AP \cdot AQ} \cdot \frac{AP \cdot AQ}{AC \cdot AF}$$

$$= \frac{\triangle ABC}{\triangle ABD} \cdot \frac{\triangle ABF}{\triangle ABE} \cdot \frac{\triangle ABD}{\triangle ABP} \cdot \frac{1}{1} \cdot \frac{\triangle ABP}{\triangle ABF} \cdot \frac{\triangle ABQ}{\triangle ABC}$$

$$= 1$$

Ex7.8 and Ex7.9 adequately shows the effect of mutual transformation between the area ratio and the line segment ratio. And both of them have made use of the transition technique: in Ex7.8, $\triangle AMC$ plays the role of transition since it can compose co-angle triangles with $\triangle ACP$ and $\triangle BCP$ respectively. In Ex7.9, $\triangle ACF$ and $\triangle BDE$ are correlated by means of the transition of $\triangle BPQ$ and $\triangle APQ$.

The following is a distinctive problem:

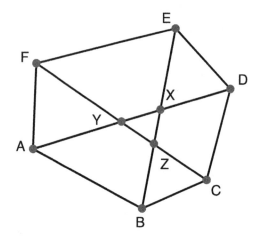

FIGURE 7.9

[Ex7.10] (Polish Mathematics Competition, 1965 - 1966) Convex hexagon *ABCDEF* has three main diagonal lines *AD*, *BE* and *CF,* all of which bisect the area. **Prove:** *AD*, *BE* and *CF* intersect at one point.

Proof: Prove by contradiction. Suppose that AD, BE and CF do not intersect at one point. Denote X as the intersection point of AD and BE, Y as the intersection point of AD and CF, and Z as the intersection point of CF and BE, as shown in Fig. 7.9.

From the given, S_{ABCF} and S_{ABCD} are both half of this hexagon's area. So,

$$S_{ABCF} = S_{ABCD}$$

Subtracting S_{ABCY} on both sides of the equation above, we have:

$$\triangle AYF = \triangle DYC$$

Likewise, we get $\triangle AXB = \triangle DXE$ and $\triangle FZE = \triangle BZC$. Hence,

$$
\begin{aligned}
1 &= \frac{\triangle AXB}{\triangle DXE} \cdot \frac{\triangle DYC}{\triangle AYF} \cdot \frac{\triangle FZE}{\triangle BZC} \\[2mm]
&= \frac{AX \cdot BX}{DX \cdot EX} \cdot \frac{DY \cdot CY}{AY \cdot FY} \cdot \frac{EZ \cdot FZ}{BZ \cdot CZ} \\[2mm]
&= \frac{(AY + XY)}{AY} \cdot \frac{(BZ + XZ)}{BZ} \cdot \frac{(DX + XY)}{DX} \cdot \frac{(CZ + ZY)}{CZ} \cdot \frac{(EX + XZ)}{EX} \cdot \frac{(FY + YZ)}{FY} > 1
\end{aligned}
$$

This is a contradiction.

ADDITIONAL PROBLEMS

[P7.1] In $\triangle ABC$, an exterior angle bisector AE intersects the extension line of BC at point E.

Prove:

$$\frac{BE}{CE} = \frac{AB}{AC}$$

.

[P7.2] Apply the Co-Angle Theorem to prove: in any triangle, a line segment connecting the midpoints of two sides is equal to half of the third side.

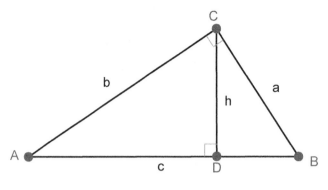

FIGURE 7.10

[P7.3] Given: in right triangle *ABC*, *CD* is the altitude of the hypotenuse, as shown in Fig.7.10.

Prove:

$(1) h = \dfrac{ab}{c}$;

$(2) \dfrac{c}{h} = \dfrac{b}{a} + \dfrac{a}{b}$;

$(3) c^2 = a^2 + b^2$;

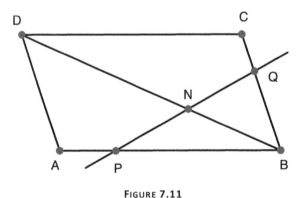

FIGURE 7.11

[P7.4] In $\square_{ABCD}$, a line intersects *AB* and *BC* at points *P* and *Q* respectively. Given: *BP* = 3*PA*, *BQ* = 2*QC* and *PQ* intersects diagonal line *BD* at point *N*. Determine the ratio of *BN* to *BD* (*fig. 7-11*).

64

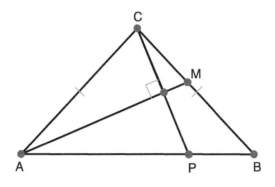

FIGURE 7.12

[P7.5] Given: in isosceles right triangle *ABC*, *M* is the midpoint of *BC*. Draw a perpendicular to *AM* from *C*, intersecting *AB* at *P*. Prove: *AP* = 2*PB (fig. 7.12)*.

[P7.6] In Ex7.7, *M* is a trisection point of *BC* instead of the midpoint, that is *BM* = 2*MC*. If $\frac{AB}{AE}$ and $\frac{AC}{AF}$ are given, calculate $\frac{AM}{AN}$.

[P7.7] Determine Ex7.7 by the Point of Division formula.

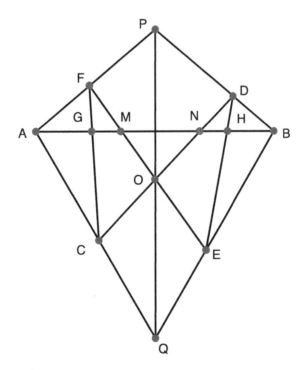

FIGURE 7.13

[P7.8] Given: PQ is the perpendicular bisector of AB. O is any point on PQ. Draw a line through O such that it intersects AP, AB and BQ at F, M and E. Draw another line through O such that it intersects AQ, AB and BP at C, N and D. CF and DE intersect AB at G, H respectively, as shown in Fig.7.13.

Prove:

$$\frac{MG}{AG} = \frac{NH}{BH} = \frac{MB}{NA}$$

CHAPTER 8

THINK FROM THE CONTRARY AGAIN

Co-Angle Inequality

The condition of the Co-Side Theorem is that two lines intersect. Think from the contrary: what if they do not? It leads us to come up with the relationship between co-side triangles and parallel lines, which works out fine.

The condition of Co-Angle Theorem is that two angles are equal or supplementary. Think from the contrary: what if the two angles are neither equal nor supplementary?

This thought does make sense, which yields an important proposition:

Co-Angle Inequality: If$\angle ABC > \angle A'B'C'$, and their sum is less than 180°, then

$$\frac{\triangle ABC}{\triangle A'B'C'} > \frac{AB \cdot BC}{A'B' \cdot B'C'}$$

alternatively:

$$\frac{\triangle ABC}{AB \cdot BC} > \frac{\triangle A'B'C'}{A'B' \cdot B'C'}$$

Proof: Let $\triangle ABC = \alpha, \angle A'B'C' = \beta$.

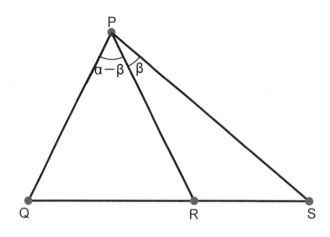

FIGURE 8.1

As shown in Fig.8.1: draw an isosceles triangle $\triangle PQR$ whose vertical angle is $\alpha - \beta$. Extend line QR to point S such that $\angle RPS = \beta$. So, $\angle QPS = \alpha$. By the Co-Angle Theorem, we have

$$\frac{\triangle ABC}{AB \cdot BC} = \frac{\triangle QPS}{PQ \cdot PS} > \frac{\triangle RPS}{PR \cdot PS} = \frac{\triangle A'B'C'}{A'B' \cdot B'C'}$$

Please think for a while: how is the condition, $\angle ABC + \angle A'B'C' < 180°$, used in the proof? (Hint: Why can we extend the line QR to S such that $\angle RPS = \beta$?)

With the Co-Angle Inequality, a series of basic geometry inequalities can be produced.

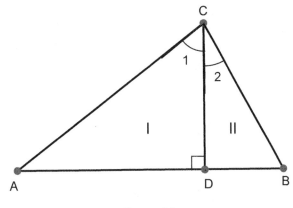

FIGURE 8.2

[Ex8.1] In ∆ABC, if ∠B > ∠C, AC > AB. Try proving it.

Proof: As shown in Fig. 8.2, take ∆ABC and ∆ACB as two triangles. Applying the Co-Angle Inequality to them, we have

$$I = \frac{\triangle ABC}{\triangle ACB} > \frac{AB \cdot BC}{AC \cdot BC} = \frac{AB}{AC}$$

$$\therefore AC > AB$$

Ex8.1 can be stated as: in any triangle, the bigger angle corresponds with the longer side. And since we have proved that the equal angles correspond with equal sides (Ex7.1), it can be inferred that the longer side corresponds with the bigger angle, which can be proved by contradiction:

If *AC* > *AB* while ∠C is no bigger than ∠C, there are two possibilities:

(1)∠B = ∠C, which definitely results in *AC* = *AB*, a contradiction to the conditions.

(2)∠B < ∠C, since the bigger angle corresponds with the longer side, we have *AB* > *AC*, a contradiction to the conditions. It definitely indicates that ∠B > ∠C

[Ex8.2] In any triangle, a sum of any two sides is bigger than the third one. Try Proving it.

Proof: Since the bigger angle corresponds to the longer side, it is sufficient to prove the sum of two opposite sides of the smaller angles are larger than that of the biggest angle. As shown in Fig.8.2, in ∆ABC, let ∠ACB be greater than or equal to ∠A and ∠B. *CD* is the altitude of ∆*ABC*. By the Co-Angle Inequality, we have

$$I = \frac{\triangle ADC}{\triangle ACD} > \frac{AD \cdot DC}{AC \cdot DC} = \frac{AD}{AC}$$

that is AC > AD. Likewise, BC > BD.

Hence

$$AC + BC > AD + BD = AB$$

[Ex8.3] In △ABC, take point P on side BC.

Prove: the length of AP is less than that of the longer of AB and AC.

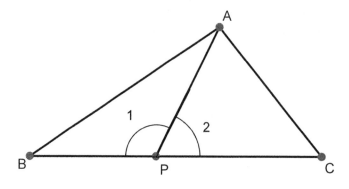

FIGURE 8.3

Proof: As shown in Fig. 8.3, since $\angle B + \angle C < 180° = \angle 1 + \angle 2$ one of the following definitely holds:

$$\angle B < \angle 1 \text{ and } \angle C < \angle 2.$$

Hence, either one of AP < AB and AP < AC definitely holds.

[Ex8.4] In △ABC and △ A′B′C′, AB = A′B′, BC = B′C′, yet $\angle ABC > \angle A'B'C'$

Prove: $AC > A'C'$

Proof: As shown in Fig.8.4, merge the two triangles by coinciding AB and A′B′. Let C and C′ lie on the same side of AB.

By BC = B′C′, we have

$$\angle BC'C = \angle BCC'$$

$$\therefore \angle AC'C = \angle BC'C + \angle AC'B > \angle BCC' - \angle ACB = \angle ACC'$$

$$\therefore AC > AC' = A'C'$$

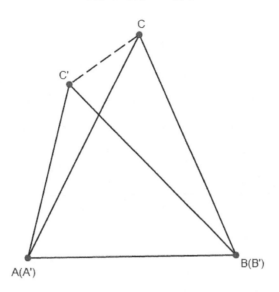

FIGURE 8.4

Ex8.4 shows: in ΔABC, we suppose the lengths of *AB* and *BC* are fixed in ΔABC. If ∠ B varies, *AC* becomes longer when ∠ B grows and *AC* becomes shorter when ∠ B shrinks. This is in accordance with our intuition.

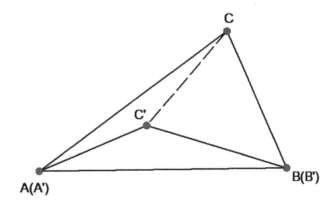

FIGURE 8.5

Attentive readers may find that the reasoning process in the proof of Ex8.4 depends on the location of C' in Fig.8.4. If Fig.8.4 changes into Fig.8.5, the proof should be adjusted to the following:

$$BC = B'C'$$

$$\therefore \angle BC'C = \angle BCC'$$

$$\therefore \angle AC'C = (\angle AC'C + \angle BC'C) - \angle BC'C > \angle ACB - \angle BCC' = \angle ACC'$$

$$\therefore AC > AC' = A'C'$$

This proof makes use of the condition that $\angle AC'C + \angle BC'C > \angle ACB$. It's obvious since C is in the interior of $\triangle ABC$, which means

$\angle AC'C + \angle BC'C > 180°$ whereas $\angle ACB < 180°$.

However, it is troublesome to have to consider two configurations. Can we avoid such trouble? We can indeed! The solution is to try to make sure C' is definitely in the exterior of $\triangle ABC$ as in Fig.8.4. It is sufficient to align the shorter side of AB and BC with that of A'B' and B'C' when merging the two triangles $\triangle ABC$ and $\triangle A'B'C'$. According to Ex8.3, it is then assured that C' must fall in the exterior of $\triangle ABC$. The reason is left for you to ponder.

The following reasoning is used in the proof of Ex8.4:

$$from\ BC = B'C',\ we\ have\ \angle BCC' = \angle BC'C$$

namely the property that the two base angles are equal in isosceles triangles. Can this be deduced by the method of comparing areas? Indeed:

[Ex8.5] In $\triangle ABC$, $AB = AC$.

Prove: $\angle B = \angle C$.

Proof: Prove by contradiction. Suppose $\angle B > \angle C$, by the Co-Angle Inequality, and we have:

$$I = \frac{\triangle ABC}{\triangle ACB} > \frac{AB \cdot BC}{AC \cdot BC} = \frac{AB}{AC}$$

Thus, $AC > AB$, which is contradictory to the condition.

Maybe you think the examples above are too simple. In fact, all the complex things consist of the simple things. If you have mastered those simple ones, the complex ones will no longer be difficult. Various kinds of inequalities in geometry can almost be deduced from the inequalities above. Ultimately, all of them can be derived from Co-Angle Inequality.

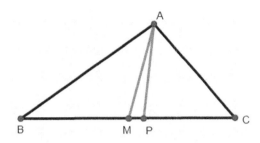

FIGURE 8.6

[Ex8.6] **Prove:** The median of a triangle is not shorter than the corresponding angle bisector. **Proof:** As shown in Fig.8.6, AM is the median of $\triangle ABC$, and AP is the angle bisector. Let's suppose $AB \geq AC$, that is $\angle C \geq \angle B$. By Co-Angle Theorem, we have

$$\frac{BP}{PC} = \frac{\triangle ABP}{\triangle ACP} = \frac{AB \cdot AP}{AC \cdot AP} = \frac{AB}{AC} \geq 1$$

Point M is the midpoint of BC, so P lies on the line segment MC (at most it coincides with M). Thus

$\angle PAC \leq \angle MAC$

$\therefore \angle AMP = \angle B + \angle BAM \leq \angle B + \angle BAP \leq \angle C + \angle PAC = \angle MPA$

$\therefore AP \leq AM$ （the bigger angle corresponds with the longer side）

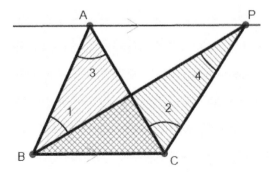

FIGURE 8.7

[Ex8.7] Given: In $\triangle ABC$, $\angle A \leq 90°$ and $AB = AC$. Draw a line AP through point A, which is parallel to BC.

Prove: $AB \cdot AC < PB \cdot PC$.

Analysis: As shown in Fig.8.7, it is sufficient to prove $\angle 4 < \angle 3$. By Co-Angle Inequality, we have:

$$1 = \frac{\triangle ABC}{\triangle PBC} > \frac{AB \cdot AC}{PB \cdot PC}$$

Thus, the problem has been solved.

To prove $\angle 4 < \angle 3$, we may first prove $\angle 2 > \angle 1$. Since

$$\angle 1 + \angle 3 = \angle 2 + \angle 4 \text{ (why?)}$$

So, is it assured that $\angle 2 > \angle 1$? Prove by contradiction. Suppose $\angle 2 < \angle 1$. By Co-Angle Inequality, we have

$$1 = \frac{\triangle ABP}{\triangle ACP} \geq \frac{AB \cdot AP}{AC \cdot CP} = \frac{BP}{CP},$$

which means $CP \geq BP$, that is $\angle CBP \geq \angle BCP$. It is impossible!

Reversing the analysis above, one can write the following proof:

Proof: *Since* $AB = AC$, *we have* $\angle ABC = \angle ACB$.

$$\therefore \angle PBC = \angle ABC - \angle 1 < \angle ACB + \angle 2 = \angle PCB$$

$$\therefore PC < PB \tag{8.1}$$

Then, let's prove $\angle 1 < \angle 2$. Prove by contradiction. Suppose $\angle 1 \geq \angle 2$. By Co-Angle Inequality and that $AP \parallel BC$, we have

$$1 = \frac{\triangle ABP}{\triangle ACP} \geq \frac{AB \cdot AP}{AC \cdot CP} = \frac{BP}{CP}$$

$$\therefore PC \geq PB \tag{8.2}$$

It is contradictory to (8.1).

So $\angle 1 < \angle 2$, $\therefore \angle 3 > \angle 4$

By Co-Angle Inequality again, we have (since$\angle BAC \leq 90°$, and $\angle BAC + \angle BPC < 180°$)

$$1 = \frac{\triangle ABC}{\triangle PBC} > \frac{AB \cdot AC}{PB \cdot PC}$$

$$PB \cdot PC > AB \cdot AC$$

Derived from Ex8.7, when its area and the length of one side are determined, the perimeter of a triangle is the shortest when its other two sides are equal. This problem is left as an exercise for you.

The following example is the famous **Steiner—Remios Theorem**.

It is easy to prove that the angle bisectors of base angles are equal in an isosceles triangle. As early as more than 2000 years ago, Euclid elaborated the theorem in his book—*Euclid's Elements*. It is easy to prove this theorem by Co-Angle Theorem. As shown in Fig.8.8, in $\triangle ABC$, if $AB = AC$, then $\angle ABC = \angle ACB$. Suppose BP and CQ are two angle bisectors, so$\angle 1 = \angle 2$. By Co-Angle Theorem, we have

$$\frac{\triangle PBC}{\triangle QBC} = \frac{PC \cdot BC}{QB \cdot BC} = \frac{PC \cdot PB}{QB \cdot QC}$$

Immediately we find that $PB = QC$ after reduction.

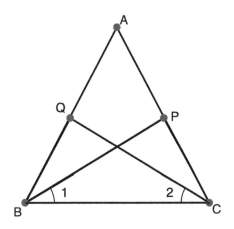

FIGURE 8.8

Inversely, in △ABC, if the two angle bisectors of ∠B and ∠C are equal, is it definite that AC = AB? It seems correct, yet Euclid failed to give the proof of the proposition. 2000 years later, Remious, a mathematician in the 18th century, specifically pointed it out as such a problem that looks simple but is hard to find a starting point. As a response to Remious, the famous geometrician Steiner gave the proof. 100 years later, hundreds of proofs to this theorem have been put forward by other people.

By Co-Angle Inequality, we can obtain a very simple proof.

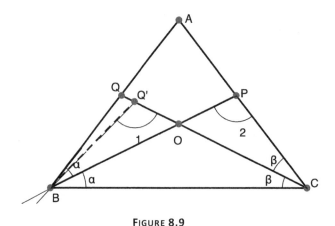

FIGURE 8.9

[Ex8.8] Given: in △ABC, BP is the angle bisector of ∠ABC, CQ is the angle bisector of ∠ACB and BP = CQ.

Prove: AB = AC (***Steiner—Remious Theorem***).

Proof: Suppose $AC \geq AB$, prove $AC > AB$ is invalid.

$$\angle BQ'C = \angle CPB, \qquad \angle Q'BC \geq PCB$$

As shown in Fig.8.9 $\alpha = \dfrac{1}{2}\angle B, \quad \beta = \dfrac{1}{2}\angle C.$ When $AC \geq AB$ it's definite that $\alpha \geq \beta$. So we can take point Q' on QO such that $\angle Q'BP = \beta$. Then, by the Co-Angle Theorem and the Co-Angle Inequality, we have:

$$\frac{BQ' \cdot CQ}{PC \cdot PB} = \frac{\triangle Q'BC}{\triangle PCB} \geq \frac{BQ' \cdot BC}{PC \cdot BC}$$

$$\therefore \frac{CQ'}{PB} \geq 1$$

that is,

$$CQ' \geq PB = QC$$

But we've known that $CQ' \leq QC$ when drawing the figure, so $CQ' = QC$, that is Q' coincides with Q. Thus, $\alpha = \beta$.

ADDITIONAL PROBLEMS

[P8.1] Prove the following conclusion with Co-Angle Inequality:

if $\angle ABC > \angle A'B'C'$ and $\angle ABC + \angle A'B'C' > 180°$, then

$$\frac{\triangle ABC}{\triangle A'B'C'} < \frac{AB \cdot BC}{A'B' \cdot B'C'}$$

[P8.2] If the above inequality holds, is it definite that $\angle ABC > \angle A'B'C'$?

[P8.3] Prove: The sum of any two sides of a triangle is bigger than twice the median of the third side.

[P8.4] Prove: In triangles whose area and the length of one side are determined, the isosceles triangle whose base is the given side, has the shortest perimeter.

[P8.5] Given: PQ ∥ AB. M is the midpoint of AB, PM > QM, and $\angle AQB \leq 90°$. Prove:
$$PA \cdot PB > QA \cdot QB$$

[P8.6] In $\triangle ABC$, take point Q, P on side AB, AC respectively such that

$$\angle 1 = \frac{1}{3}\angle ABC, \angle 2 = \frac{1}{3}\angle ACB$$

as shown in Fig.8.8.

If $BP=CQ$. Prove $AB=AC$.

CHAPTER 9

THINK CONVERSELY

Co-Angle Converse Theorem

Thinking about the contrary of the Co-Angle Theorem, we come up with the Co-Angle Inequality.

Thinking about a problem, we may not only think from the contrary, but also conversely. The Co-Angle Theorem says that if $\angle ABC$ and $\angle A'B'C'$ are equal or supplementary, the

equation $\dfrac{\triangle ABC}{\triangle A'B'C'} = \dfrac{AB \cdot BC}{A'B' \cdot B'C'}$ holds.

Conversely, if $\dfrac{\triangle ABC}{\triangle A'B'C'} = \dfrac{AB \cdot BC}{A'B' \cdot B'C'}$ holds, is it definite that two angles, $\angle ABC$

and $\angle A'B'C'$ are equal or supplementary?

By the Co-Angle Inequality, it's easy to solve this question. The answer is yes, and we call it the Co-Angle Converse Theorem.

Co-Angle Converse Theorem: If the area ratio of $\triangle ABC$ and $\triangle A'B'C'$ is equal to the product ratio of two angular sides about $\angle B$ and $\angle B'$, that is

$$\frac{\triangle ABC}{\triangle A'B'C'} = \frac{AB \cdot BC}{A'B' \cdot B'C'}$$

then $\angle B$ and $\angle B'$ are equal or supplementary.

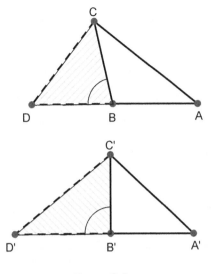

FIGURE 9.1

Proof: Prove by contradiction. If $\angle B$ and $\angle B'$ are neither equal nor supplementary, we may set $\angle B > \angle B'$. Now, there are two cases:

$\angle B + \angle B' < 180°$ or $\angle B + \angle B' > 180°$.

If $\angle B + \angle B' < 180°$, by the Co-Angle Inequality, we have

$$\frac{\triangle ABC}{\triangle A'B'C'} > \frac{AB \cdot BC}{A'B' \cdot B'C'}$$

It is contradictory to the given condition.

If $\angle B + \angle B' > 180°$, as shown in Fig.9.1, extend AB to D such that $BD=AB$, and extend $A'B'$ to D' such that $B'D' = A'B'$.

Then, $\angle DBC + \angle D'B'C' < 180°$ and $\angle DBC = 180 - \angle B < 180° - \angle B' = \angle D'B'C'$.

By the Co-Angle Inequality, we have

$$\frac{\triangle D'B'C'}{\triangle DBC} > \frac{B'D' \cdot B'C'}{BD \cdot BC}$$

But since $\triangle D'B'C' = \triangle A'B'C'$, $\triangle DBC = \triangle ABC$, $B'D' = A'B'$, and $BD=AB$, the equation above can be rewritten as:

$$\frac{\triangle ABC}{\triangle A'B'C'} < \frac{AB \cdot BC}{A'B' \cdot B'C'}$$

which is contradictory to the given condition.

By the Co-Angle Converse Theorem, we may prove the equality or supplementarity of two angles. Let's first look at two simple examples.

[Ex9.1] In $\triangle ABC$, if $AB = AC$, then $\angle B = \angle C$.

Proof: According to the given, we have

$$\frac{\triangle ABC}{\triangle ACB} = \frac{1}{1} = \frac{AB \cdot BC}{AC \cdot BC}$$

By Co-Angle Converse Theorem, $\angle B$ and $\angle C$ are equal or supplementary. Since $\angle B + \angle C < 180°$ they can't be supplementary. Thus, they are equal. □

The Co-Angle Converse Theorem gives us another method of proving Ex8.5.

[Ex9.2] Given: in $\triangle ABC$ and $\triangle A'B'C'$, $\angle A = \angle A'$ and

$$\frac{AC}{A'C'} = \frac{BC}{B'C'}$$

Then, when $\angle B$ and $\angle B'$ are not supplementary, we have

$$\triangle ABC \sim \triangle A'B'C'$$

Proof: Since $\angle A = \angle A'$, by the Co-Angle Theorem, along with the assumption, we have

$$\frac{\triangle ABC}{\triangle A'B'C'} = \frac{AB \cdot AC}{A'B' \cdot A'C'} = \frac{AB \cdot BC}{A'B' \cdot B'C'}$$

By the Co-Angle Converse Theorem, we know that $\angle B$ and $\angle B'$ are equal or supplementary. If $\angle B$ and $\angle B'$ are not supplementary, this means $\angle B$ equals $\angle B'$. We know that

$\triangle ABC \sim \triangle A'B'C'$ since $\angle B = \angle B'$ and $\angle A = \angle A'$ (the conclusion of Ex7.3 is used here).

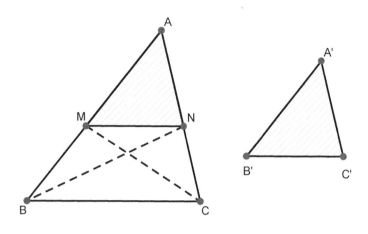

FIGURE 9.2

[Ex9.3] If the corresponding sides of $\triangle$ ABC and $\triangle$ $A'B'C'$ are proportional, then the triangles are similar.

Proof: Suppose $\triangle ABC$ is bigger, and take point M and N on side AB and AC respectively such that $AM=A'B'$, $AN=A'C'$.

As shown in Fig.9.2, our approach is:

Step one: **Prove:** $\triangle AMN \sim \triangle ABC$

Step two: **Prove:** $\triangle AMN \cong \triangle A'B'C'$, thus $\triangle ABC \sim \triangle A'B'C'$.

By the given, we have

$$\frac{AM}{AB} = \frac{A'B'}{AB} = \frac{A'C'}{AC} = \frac{AN}{AC}$$

$$\therefore \frac{\triangle AMC}{\triangle AMN} = \frac{AC}{AN} = \frac{AB}{AM} = \frac{\triangle ANB}{\triangle AMN}$$

$$\therefore \triangle AMC = \triangle ANB$$

$$\therefore \triangle BMN = \triangle CMN$$

$$\therefore BC \parallel MN$$

$$\therefore \angle ABC = \angle AMN$$

And since $\angle MAN = \angle BAC$, $\triangle AMN \sim \triangle ABC$. Step one is proved.

By $\triangle AMN \sim \triangle ABC$, we have

$$\frac{MN}{BC} = \frac{AN}{AC} = \frac{A'C'}{AC} = \frac{B'C'}{BC}$$

$$\therefore MN = B'C'$$

$$\therefore \triangle AMN \cong \triangle A'B'C'$$

$$\therefore \triangle ABC \sim \triangle A'B'C'$$

In the last step, the **SSS Congruence Theorem** is used. This criterion can be inferred from Ex8.4, which has already been proved. Just by contradiction, supposing that $\angle AMN > \angle A'B'C'$ (or $\angle AMN < \angle A'B'C'$). From Ex8.4, we know that in this case, when $AM = A'B'$ and $MN = B'C'$, then $AN > A'C'$ (or $AN < A'C'$) definitely holds, which is a contradiction.

Hence, $\angle AMN = \angle A'B'C'$.

[Ex9.4] (ASS Congruence Theorem) In $\triangle ABC$ and $\triangle A'B'C'$, if $\angle A = \angle A'$, $AB = A'B'$, $BC = B'C'$, and $\angle C$ and $\angle C'$ aren't supplementary, then $\triangle ABC \cong \triangle A'B'C'$. Try proving it.

Proof: By the Co-Angle Theorem and the given, we have

$$\frac{\triangle ABC}{\triangle A'B'C'} = \frac{AB \cdot AC}{A'B' \cdot A'C'} = \frac{AC}{A'C'} = \frac{AC \cdot BC}{A'B' \cdot B'C'}$$

By the Co-Angle Converse Theorem, we know $\angle C$ and $\angle C'$ are equal or supplementary. But it is given that $\angle C$ and $\angle C'$ aren't supplementary, thus $\angle C$ equals $\angle C'$. Hence, $\triangle ABC \cong \triangle A'B'C'$ is derived.

In geometry curriculums, the criteria for congruent triangles include SAS, ASA, AAS, SSS but not ASS. This is just because the criterion for ASS requires an auxiliary condition that opposite angles of a pair of corresponding sides are not supplementary. Quite a few geometry theorems are valid under certain auxiliary conditions. These auxiliary conditions, called non-degeneracy conditions, are generally inequalities. It is a contribution of Chinese mathematician Wu Wenjun to point out the importance of non-degeneracy conditions and demonstrate how to determine the non-degeneracy conditions which pertain.

For a long time, it was thought that the traditional methods of geometry proof are very rigorous and the reasoning of geometry proofs was regarded as the model of logical reasoning. But, by researching machine proof of geometry theorems, Professor Wu found an important fact: the traditional methods of geometry proofs are not only not rigorous but also unable to admit a truly rigorous proof. This is because the validity of geometry theorems often requires non-degeneracy conditions, whereas the traditional methods are unable to determine these non-degeneracy conditions. Especially for complex theorems, it is not easy to determine non-degeneracy conditions. Professor Wu established an internationally accepted method of machine proof of geometry theorems, which is called **Wu's Method**. By Wu's Method, one may prove all the geometry theorems of equality type on a computer, and can even find new theorems. Such a method of machine proof is very rigorous and is able to determine the non-degeneracy conditions which make a theorem valid.

Let's return to the Co-Angle Converse Theorem and use it to prove two nontrivial propositions.

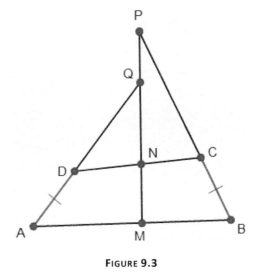

FIGURE 9.3

[Ex9.5] In quadrilateral $ABCD$, M and N are the midpoints of side AB and CD, respectively, while the other two sides satisfy $AD = BC$. Extend side BC and AD such that they intersect line MN at P and Q, respectively (as shown in Fig.9.3).

Prove: $\angle AQM = \angle BPM$.

Proof: Since M and N are the midpoints of AD and BC respectively, we have

$$\frac{AQ}{DQ} = \frac{\triangle AMN}{\triangle DMN} = \frac{\triangle BMN}{\triangle CMN} = \frac{BP}{CP}$$

that is

$$\frac{AQ}{AQ - AD} = \frac{BP}{BP - BC}$$

With $AD = BC$, we have $AQ = BP$.

$$\therefore \frac{\triangle AMQ}{\triangle PMB} = \frac{AM \cdot QM}{BM \cdot PM} = \frac{QM}{PM} = \frac{AQ \cdot QM}{BP \cdot PM}$$

By the Co-Angle Converse Theorem, we know that $\angle AQM$ and $\angle BPM$ are equal or supplementary. But they are obviously not supplementary ($\because \angle A + \angle B + \angle AQM + \angle BPM = 180°$). Hence $\angle AQM = \angle BPM$.

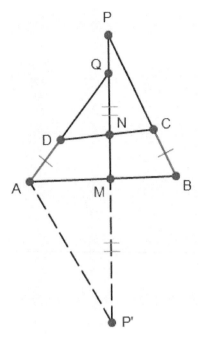

FIGURE 9.4

There is a smart method for this example. As shown in Fig.9-4, extend PM to P' such that $P'M = PM$. Then $\triangle AMP' \cong \triangle BMP, \angle P' = \angle P$. Yet by the proved condition that $AQ=BP=AP'$, we know that $\triangle P'AQ$ is an isosceles triangle. Thus $\angle P' = \angle AQM$. The proof has been completed.

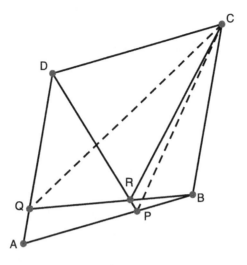

FIGURE 9.5

[Ex9.6] In parallelogram $ABCD$, take point Q on AD and point P on AB.

Suppose $BQ = DP$ and that DP and BQ intersect at R.

Prove: $\angle DRC = \angle BRC$.

Proof: As shown in Fig.9.5, since $AB \parallel DC$, we have $\triangle DPC = \triangle DBC$.

since $AD \parallel BC$, we have

$$\triangle QBC = \triangle DBC$$
$$\therefore \triangle PDC = \triangle QBC$$

$$\frac{\triangle DRC}{\triangle BRC} = \frac{\triangle DRC}{\triangle PDC} = \frac{\triangle QBC}{\triangle BRC}$$

$$= \frac{DR}{DP} \cdot \frac{BQ}{BR} = \frac{DR}{BR}$$

$$= \frac{DR \cdot RC}{BR \cdot RC}$$

By the Co-Angle Converse Theorem, $\angle$DRC and $\angle$BRC are equal or supplementary. Since it is impossible for them to be supplementary, they are equal.

ADDITIONAL PROBLEMS

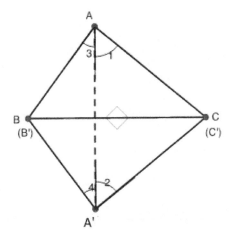

FIGURE 9.6

[P9.1] A student "proved" the ASS Congruence Theorem using the following method:

Given: In $\triangle ABC$ and $\triangle A'B'C'$, $\angle A = \angle A'$, $BC = B'C'$ and $AC = A'C'$

Prove: $\triangle ABC \cong \triangle A'B'C'$.

Proof: Merge two triangles by coinciding side BC and $B'C'$, as shown in Fig.9.6.

$\because AC = A'C'$ (given)

$\therefore \angle 1 = \angle 2$

$\because \angle BAC = \angle B'A'C' = \angle BA'C$ (given)

$\therefore \angle 3 = \angle BAC - \angle 1$

$= \angle B'A'C' - \angle 2$

$= \angle 4$

$\therefore AB = A'B = A'B'$

$\therefore \triangle ABC \cong \triangle A'B'C'$ (SSS)

Can you point out the faults of the "proof"?

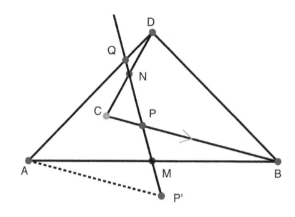

FIGURE 9.7

[P9.2] In isosceles triangle $\triangle ABD$, M is the midpoint of base AB. Take point C in the interior of $\triangle ABD$ such that $BC = BD$. Take the midpoint N on CD and connect M and N such that MN intersects AD at point Q and BC at P.

Prove: $\angle DQN = \angle BPN$.

Chapter 10

Area Equations

To solve geometry problems, one can absorb algebraic thoughts and methods. Previously, we absorbed the thought of elimination in algebra and introduced the "Eliminating Points Method", which brought us lots of benefits. Methods of formulation in algebra are also worth our consideration.

It is pretty easy to formulate some equalities by making use of the relationships between geometric quantities. If these equalities include indeterminate variables, then they take the form of algebraic equations.

Using different methods to calculate the same area should lead to the same result, and thus yield an equation. The equations formulated by using the relationship of area equality are called *Area Equations*.

The following are simple applications of Area Equations.

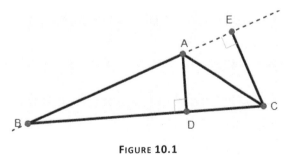

FIGURE 10.1

[Ex10.1] Given: AD and CE are the altitudes of $\triangle ABC$ and $AD = 5$cm, $CE = 7$cm, and $AB = 15$cm. Determine the length of BC (as shown in Fig. 10.1).

Solution: Use two methods to calculate the area of $\triangle ABC$:

$$\frac{1}{2}AD \cdot BC = \triangle ABC = \frac{1}{2}CE \cdot AB$$

Substituting the known quantities into the equation above, we have

$$\frac{1}{2} \times 5 \times BC = \frac{1}{2} \times 7 \times 15$$

It's solved for $BC = 21$(cm).

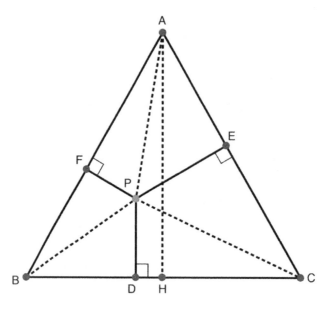

FIGURE 10.2

[Ex10.2] **Given:** $\triangle ABC$ is an equilateral triangle, and its height $h = 4$cm. The distance from point P to AB and to AC is 1cm and 2cm respectively. Determine the distance from point P to BC.

Solution: As shown in Fig.10.2, let AH be the altitude of $\triangle ABC$. The perpendicular line segments from P to BC, AC and AB are PD, PE and PF respectively. The side length of $\triangle ABC$ is a. By the relationship of areas, we have:

$$\triangle PAB + \triangle PBC + \triangle PCA = \triangle ABC$$

$$\frac{1}{2}a \cdot PF + \frac{1}{2}a \cdot PD + \frac{1}{2}a \cdot PE = \frac{1}{2}a \cdot AH$$

Substituting the known quantity, and dividing the two sides by $\frac{1}{2}a$ we have

$$1 + PD + 2 = 4$$

It's solved for $PD = 1$.

Think about it: is such a solution complete?

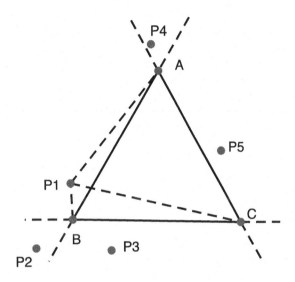

FIGURE 10.3

Attentive readers will find that we add a condition that P is in the interior of $\triangle ABC$ in Fig.10.2 implicitly, which is not in the given. If P is not in the interior of $\triangle ABC$ as shown in Fig.10.3, the position of P_1 , we formulate the Area Equation as follows:

$$\triangle P_1AC + \triangle P_1BC - \triangle P_1AB = \triangle ABC$$

After reduction, we have $2 + P_1D - 1 = 4$, then $P_1D = 3$.

If we take the position of P_2 in Fig.10.3, the equation becomes:

$$\triangle P_2AC - \triangle P_2AB - \triangle P_2BC = \triangle ABC$$

After reduction, we have $2 - 1 - P_2D = 4$, which is solved for $P_2D = -3$. This means that P cannot be in position P_2.

Similar analysis yields that if point P is not in the interior of $\triangle ABC$, it can be at positions P_1 , P_4 , P_5 and so on in Fig.10.3. When P lies at P_4 , the Area Equation is

$$\triangle P_4BC - \triangle P_4AB - \triangle P_4AC = \triangle ABC$$

After reduction, we have $P_4D - 1 - 2 = 4$, that is . $P_4D = 7$. When P lies at P_5, the Area Equation is

$$\triangle P_5AB + \triangle P_5BC - \triangle P_5AC = \triangle ABC$$

After reduction, we have $1 + P_5D - 2 = 4$, $P_5D = 5$

Therefore, the possible solutions are 1, 3, 5, 7, which respectively corresponds to P in the interior of $\triangle ABC$; P in the exterior of $\triangle ABC$ but in the interior of $\angle BCA$; P in the exterior of $\triangle ABC$ but in the interior of $\angle ABC$; P in the interior of the opposite angle of $\angle BAC$.

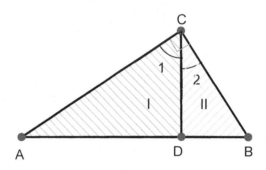

FIGURE 10.4

[Ex10.3] Given: CD is the altitude of the hypotenuse AB of a right triangle (as shown in Fig.10.4).

Prove:
$$\frac{1}{CD^2} = \frac{1}{AC^2} + \frac{1}{BC^2}$$

Proof: From the given, refer to Fig. 10.4 and formulate two equations:

$$\begin{cases} AB \cdot CD = 2 \triangle ABC = AC \cdot BC & (10.1) \\ \triangle ABC = \triangle I + \triangle II & (10.2) \end{cases}$$

Dividing both sides of (10.2) by $\triangle ABC$, notice that $\angle 1 = \angle B$, $\angle 2 = \angle A$,

By the Co-Angle Theorem, we have

92

$$1 = \frac{\triangle I + \triangle II}{\triangle ABC} = \frac{AC \cdot CD}{AB \cdot BC} + \frac{BC \cdot CD}{AB \cdot AC} \qquad (10.3)$$

Solving (10.1) for AB:

$$AB = \frac{AC \cdot BC}{CD} \qquad (10.4)$$

Substituting (10.4) into (10.3), we have

$$1 = \frac{CD^2}{BC^2} + \frac{CD^2}{AC^2}$$

Dividing both sides by CD^2, we have

$$\frac{1}{CD^2} = \frac{1}{AC^2} + \frac{1}{BC^2}$$

The problem above has a few variants. If solving CD instead of AB from (10.1), we have

$$CD = \frac{AC \cdot BC}{AB}$$

Substituting into (10.3), we have

$$1 = \frac{AC^2}{AB^2} + \frac{BC^2}{AB^2}$$

Multiplying both sides by AB^2, we obtain Pythagorean Theorem (as Ex.7.3).

If we solve the following equation from (10.1)

$$AC = \frac{AB \cdot CD}{BC}$$

Substituting it into (10.3), we have

$$1 = \frac{CD^2}{BC^2} + \frac{BC^2}{AB^2}$$

It turns into another question. Of course, we can solve BC from (10.1) and obtain another similar question.

Using Area Equations, we can solve quite complicated problems.

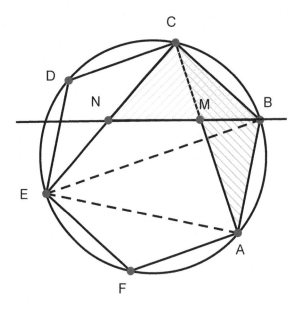

FIGURE 10.5

[Ex10.4] (International Olympics Mathematics Competition, 1982) In hexagon *ABCDEF*, diagonal lines *AC*, *CE* are respectively divided by interior points *M*, *N* into the ratio:

$$\frac{AM}{AC} = \frac{CN}{CE} = r$$

If *B*, *M*, *N* are collinear, determine *r*.

Solution: by Area Equation,

$$\triangle BCN = \triangle BCM + \triangle MCN$$

Dividing both sides by $\triangle ABC$, we have

$$\frac{\triangle BCN}{\triangle ABC} = \frac{\triangle BCM}{\triangle ABC} + \frac{\triangle MCN}{\triangle ABC}$$

By the Co-Angle Theorem, we have

$$\frac{\triangle BCN}{\triangle ABC} = \frac{\triangle BCN}{\frac{1}{2}\triangle BCE} = \frac{2BC \cdot CN}{BC \cdot CE} = 2r$$

$$\frac{\triangle BCM}{\triangle ABC} = \frac{BC \cdot CM}{BC \cdot CA} = \frac{AC - AM}{AC} = 1 - r$$

$$\frac{\triangle MCN}{\triangle ABC} = \frac{\triangle MCN}{\frac{1}{3}\triangle ACE} = \frac{3MC \cdot CN}{AC \cdot CE} = 3r(1-r)$$

Then, we obtain this equation for r

$$2r = (1 - r) + 3r(1 - r)$$

Solving for r, we have

$$3r^2 = 1$$

$$r = \pm\frac{\sqrt{3}}{3}$$

By hypothesis, $r > 0$, so

$$r = \frac{\sqrt{3}}{3}$$

.

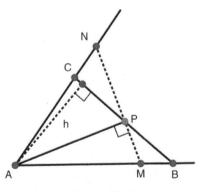

FIGURE 10.6

[Ex10.5] (United States Olympics Mathematics Competition, 1979) A fixed point P is in the interior of acute angle $\angle A$. Draw a line through P such that it intersects the two sides of $\angle A$ at B and C respectively (as shown in Fig.10.6). When will

$$\frac{1}{PB} + \frac{1}{PC}$$

attain its maximum value?

Solution: Let the height of BC in $\triangle ABC$ be h. Draw line AP perpendicular to MN which goes through P with endpoints M and N on AB and AC respectively.

By Area Equations, we have

$$\triangle APB + \triangle APC = \triangle ABC$$

Dividing both sides by $\triangle APB \cdot \triangle APC$, we have

$$\frac{1}{\triangle APC} + \frac{1}{\triangle APB} = \frac{\triangle ABC}{\triangle APB \cdot \triangle APC}$$

that is

$$\frac{2}{h \cdot PC} + \frac{2}{h \cdot PB} = \frac{\triangle ABC}{\triangle APB \cdot \triangle APC}$$

$$= \frac{\triangle ABC}{\triangle AMN} \cdot \frac{\triangle AMN}{\triangle AMP} \cdot \frac{\triangle AMP}{\triangle APB} \cdot \frac{\triangle ANP}{\triangle APC} \cdot \frac{1}{\triangle ANP}$$

$$= \frac{AB \cdot AC}{AM \cdot AN} \cdot \frac{MN}{MP} \cdot \frac{AM}{AB} \cdot \frac{AN}{AC} \cdot \frac{2}{AP \cdot PN}$$

$$= \frac{2MN}{PM \cdot PN \cdot AP}$$

$$= 2\left(\frac{1}{PM} + \frac{1}{PN}\right) \cdot \frac{1}{AP}$$

$$\therefore \frac{1}{PB} + \frac{1}{PC} = \frac{h}{AP}\left(\frac{1}{PM} + \frac{1}{PN}\right)$$

$$\leq \frac{1}{PM} + \frac{1}{PN}$$

Thus the maximum value of $\dfrac{1}{PB} + \dfrac{1}{PC}$ is $\dfrac{1}{PM} + \dfrac{1}{PN}$

and is attained when $BC \perp AP$, that is $h = AP$.

96

Area equations can be used in a variety of circumstances, and we will explore some in the following sections. First, though we'll examine some strategies in their use.

From the examples above, especially Ex10.4 and Ex10.5, you may ask: in a figure, there are several areas, thus several equations can be formulated; so how do you determine which areas should be used to formulate equations?

It seems to involve tricky problem-solving skills, but it does not mean there are no paths to follow. Three approaches of formulating equations are used in examples above:

1. Do not divide a triangle. Use different formulas to determine its area, which can help you find out the relationships between its sides, angles and altitudes. This is the simplest and most essential Area Equation. The equation (1) in both Ex10.1 and Ex10.3 follows this thought.

2. When point A, B and C are on a line, if B is between A, C, one may use any point P not on line AB to formulate an equation.

$$\triangle\,\mathrm{PAC} = \triangle\,\mathrm{PAB} + \triangle\,\mathrm{PBC}$$

This kind of technique describing three collinear points and their relationship to areas is a commonly used and effective tool. This technique is used in Ex10.3, Ex10.4 and Ex10.5.

In Ex10.4, the given indicates that B, M, N are collinear, and suggests we use the equation

$$\triangle\,BCN = \triangle\,BCM + \triangle\,MCN$$

In Ex10.5, the given indicates "draw a line through P such that it intersects the two sides of $\angle A$ at B and C", which shows P is on the line segment BC and suggests we use the Area Equation.

$$\triangle\,ABC = \triangle\,APB + \triangle\,APC$$

3. Another frequently used method to formulate Area Equations is by dividing a triangle into three parts. When the condition that three points are collinear is not particularly specified in the given, we can use this method.

When the equation has been formulated, it is worth pondering which formula should be used to determine areas in the equation. Usually, we make use of Co-Angle Theorem to transform areas into the product of line segments, especially for Area Equations formulated by the condition of collinearity. The goal of transformation is to make the geometric quantities concerned with the given and the conclusion, appear in the equation. Ex10.3, Ex10.4 and Ex10.5 all used Co-Angle Theorem, which shows the general process of solving problems by Area Equations.

ADDITIONAL PROBLEMS

[P10.1] Prove: The biggest side of a triangle corresponds with the smallest altitude.

[P10.2] Let BC be the base of the isosceles triangle $\triangle ABC$, and P is any point on BC.

> **Prove:** The sum of the distance from P to two equal sides equals the height on side AB of $\triangle ABC$.

[P10.3] Let point G be the centroid of $\triangle ABC$ (that is the intersection point of three median lines of $\triangle ABC$).

> **Prove:** The respective distance from G to three sides of $\triangle ABC$ is inversely proportional to the length of each side.

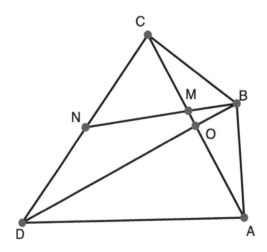

FIGURE 10.7

[P10.4] In quadrilateral $ABCD$, the diagonal line AC and BD intersect at point O. Take M, N on AC and CD respectively such that

$$\frac{AM}{AC} = \frac{CN}{CD} = k$$

If O is the midpoint of AC, how is the ratio $\dfrac{DO}{BO}$ correlated with k?

CHAPTER 11

PYTHAGOREAN DIFFERENCE THEOREM

Area equations along with the Co-Angle Theorem can solve plenty of geometry problems. In the previous section, Ex10.3 introduced two equations. Solving CD from equation (1) and substituting it into the equation (2), we obtained the **Pythagorean Theorem** immediately.

Therefore, any problem that can be solved by the Pythagorean Theorem may also be solved by Area Equations and the Co-Angle Theorem. The Pythagorean Theorem is so widely applied that it's honored as the "cornerstone of geometry". In this way, the Area Equations and Co-Angle Theorem can be regarded as the rock from which the cornerstone is quarried.

The Pythagorean Theorem is used to calculate the side length of right triangles. For general triangles, can one use the Area Equations and Co-Angle Theorem to calculate the length of sides?

We have already known that in $\triangle ABC$, if the length of two sides a, b adjacent to $\angle C$ are constant, then the larger $\angle C$, the longer its opposite side (Ex8.4). When $\angle C = 90°$, by the Pythagorean Theorem, we know $c^2 = a^2 + b^2$. Let $\angle C$ increase to $\angle ACB_2$ such that the opposite side $AB_2 > AB$, as shown in the Fig.11.1; let $\angle C$ decrease to $\angle ACB_1$ such that the opposite side $AB_1 < AB$.

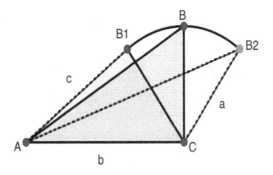

FIGURE 11.1

So, we find a law: in $\triangle ABC$, if $\angle C$ is an obtuse angle then $c^2 > a^2 + b^2$. If $\angle C$ is the acute angle then $c^2 < a^2 + b^2$.

We call $a^2 + b^2 - c^2$ **Pythagorean Difference** of $\angle C$ in $\triangle ABC$. Simply, the Pythagorean Difference of an angle of $\triangle PQR$ is $PQ^2 + QR^2 - PR^2$. We should notice that the Pythagorean Difference is correlated not only with the measure of $\angle PQR$ but also with the length of PQ and QR.

There is a very useful theorem about the Pythagorean Difference.

Pythagorean Difference Theorem: if $\angle ACB = \angle A'C'B'$ or the two angles are supplementary, then

$$\frac{a^2 + b^2 - c^2}{\triangle ABC} = \pm \frac{a'^2 + b'^2 - c'^2}{\triangle A'B'C'}$$

The choice of positive and negative sign is as follows: if the two angles are equal, then we take the positive sign; if the two angles are supplementary, we take the negative sign.

It is not too urgent to prove this theorem. What matters is to see where it can be used.

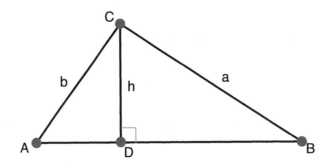

FIGURE 11.2

[Ex11.1] Given: in $\triangle\ ABC$, $BC = a$, $AC = b$, $AB = c$, determine the length h of side CD.

Solution: As shown in Fig.11.2, at least one of $\angle A$ or $\angle B$ must be acute. Let $\angle A$ be the acute angle. The altitude on AB is CD. Then,

applying Pythagorean Difference Theorem to $\triangle\ CAD\ and\ \triangle\ CAB$, we have

$$\frac{b^2 + c^2 - a^2}{\triangle\ ABC} = \frac{b^2 + AD^2 - h^2}{\triangle\ ADC}$$

Notice that $\triangle\ ABC = \frac{1}{2}ch$, $\triangle\ ADC = \frac{1}{2}AD \bullet h$, and $h^2 = b^2 - AD^2$ and substitute them into the equation above. By rearrangement, we obtain

$$\frac{b^2 + c^2 - a^2}{c} = \frac{2AD^2}{AD}$$

$$\therefore AD = \frac{1}{2c}(b^2 + c^2 - a^2)$$

$$\therefore h = \sqrt{b^2 - AD^2} = \frac{\sqrt{4b^2c^2 - (b^2 + c^2 - a^2)^2}}{2c}$$

By the way, we know that the area of $\triangle\ ABC$ is

$$\triangle\ ABC = \frac{1}{2}ch = \frac{1}{4}\sqrt{4b^2c^2 - (b^2 + c^2 - a^2)^2}$$

This formula in ancient China is called "***the Triclinic Quadrature Formula***", which is first discovered by mathematician Jiushao Qin of the Song Dynasty in Ancient China. [1]

[Ex11.2] Given: the three sides *a, b, c of* △ ABC. Determine the length of the angle bisector of ∠ACB.

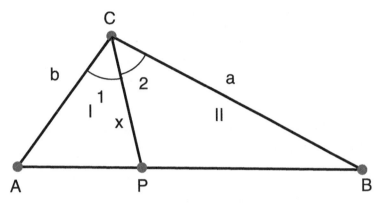

FIGURE 11.3

Solution: As shown in Fig.11.3, let *CP* be the angle bisector of ∠ACB and *CP = x.*

Applying the Pythagorean Difference Theorem to △ CAB and △ CAP, we have

$$\frac{b^2 + AP^2 - x^2}{AP} = \frac{b^2 + c^2 - a^2}{c}$$

By the property of angle bisectors,

$$\frac{AP}{BP} = \frac{b}{a}$$

that is

$$\frac{AP}{c - AP} = \frac{b}{a}, \qquad \therefore AP = \frac{bc}{a + b}$$

Substituting it into the former equation, after rearrangement, we have

---

[1] By slight rearrangement, this formula turns into ***Heron's Formula***

$$ab^2 + \frac{b^2c^2}{a+b} - (a+b)x^2 = bc^2 - a^2b$$

Solving for x:

$$x = \frac{\sqrt{ab[(a+b)^2 - c^2]}}{a+b}$$

The equation above may also be written as

$$x = \sqrt{ab\left[1 - \frac{c^2}{(a+b)^2}\right]}$$

[Ex11.3] **Given:** the length of three sides of $\triangle$ ABC are a, b and c respectively. Take point P on AB such that $AP = \frac{1}{3}c$.

Determine: CP.

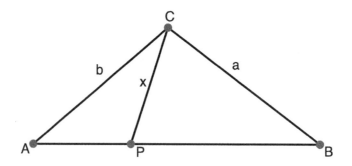

FIGURE **11.4**

Solution: As shown in Fig. 11.4, let $x = PC$. Applying Pythagorean Difference Theorem, we have

$$\frac{b^2 + AP^2 - x^2}{\triangle APC} = \frac{b^2 + c^2 - a^2}{\triangle ABC}$$

With the condition $AP = \frac{1}{3}c$, we know $\triangle APC = \frac{1}{3}\triangle ABC$.

From the equation above, we have

$$3\left[b^2 + \left(\frac{c}{3}\right)^2 - x^2\right] = b^2 + c^2 - a^2$$

Solving for x:

$$x = \frac{1}{3}\sqrt{3a^2 + 6b^2 - 2c^2}$$

[Ex11.4] Given: In parallelogram $ABCD$, $AB = c$, $BC = a$ and the diagonal line $AC = b$.

Determine: the length of the other diagonal BD.

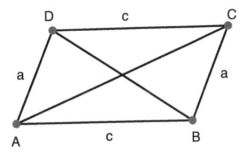

FIGURE 11.5

Solution: As shown in the Fig.11.5, apply the Pythagorean Difference Theorem to $\triangle ABC$ and $\triangle BCD$. Since $\angle ABC$ and $\angle BCD$ are supplementary, we have

$$\frac{a^2 + c^2 - b^2}{\triangle ABC} = -\frac{a^2 + c^2 - BD^2}{\triangle BCD}$$

From $\triangle ABC = \triangle BCD$, it's solved for

$$BD = \sqrt{2(a^2 + c^2) - b^2}$$

From these examples, we see that it is effective to use the Pythagorean Difference Theorem to calculate the length of line segments. For such a useful tool, we had better find out how it is derived. Next, we use the Area Equations to prove the Pythagorean Difference Theorem.

The Proof of Pythagorean Difference Theorem

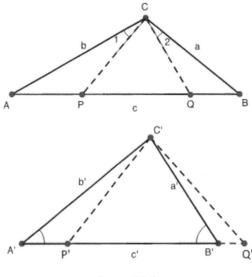

FIGURE **11.6**

First, we prove the case where two angles are supplementary. For the sake of convenience, let $\angle ACB \geq 90°$.

As shown in Fig.11.6, denote the three sides of $\triangle ABC$ as a, b, c and three sides of $\triangle A'B'C'$ as a', b' and c'.

In two figures of Fig.11.6, take points P, P' on the sides AB, $A'B'$ respectively such that

$$\angle ACP = \angle B \qquad \angle A'C'P' = \angle B'$$

Then, take points Q, Q' on the sides BA, $B'A'$ respectively such that

$$\angle BCQ = \angle A \qquad \angle B'C'Q' = \angle A'$$

It's obvious that

$$\angle CPQ = \angle CQP = \angle A + \angle B$$

$$\angle C'P'Q' = \angle C'Q'P' = 180° - (\angle A' + \angle B') = \angle A + \angle B$$

(because $\angle C$ and $\angle C'$ are supplementary).

Thus, we know PC = QC, $P'C' = Q'C'$, and $\angle PCQ = \angle P'C'Q'$

Now, write out the following two Area Equations:

$$\begin{cases} \triangle ABC = \triangle ACP + \triangle CBQ + \triangle PCQ & (11.1) \\ \triangle A'B'C' = \triangle A'C'P' + \triangle C'B'Q' - \triangle P'C'Q' & (11.2) \end{cases}$$

Dividing both sides of (11.1) by $\triangle ABC$, we have

$$1 = \frac{\triangle ACP}{\triangle ABC} + \frac{\triangle CBQ}{\triangle ABC} + \frac{\triangle PCQ}{\triangle ABC} \qquad (11.3)$$

By the Co-Angle Theorem, from $\qquad \angle ACB = \angle APC = \angle BQC = 180° - \angle QPC$

we have

$$1 = \frac{AP \bullet PC}{ab} + \frac{BQ \bullet QC}{ab} + \frac{PC \bullet PQ}{ab}$$

$$= \frac{PC(AP + BQ + PQ)}{ab} = \frac{PC \bullet c}{ab} \qquad (11.4)$$

On the other hand, applying the Co-Angle Theorem to $\angle ACP = \angle B$ and $\angle BCQ = \angle A$ in

(11.3), we have:

$$1 = \frac{b \bullet PC}{ac} + \frac{a \bullet QC}{bc} + \frac{\triangle PCQ}{\triangle ABC} \qquad (11.5)$$

From (11.4) we get: $\qquad PC = \frac{ab}{c}$

Substituting it into (11.5), we have (notice that $PC = QC$).

$$1 = \frac{b^2}{c^2} + \frac{a^2}{c^2} + \frac{\triangle PCQ}{\triangle ABC} \qquad (11.6)$$

by rearrangement of (11.6), we have

$$-\frac{c^2 \triangle PCQ}{\triangle ABC} = a^2 + b^2 - c^2 \qquad (11.7)$$

By (11.2), likewise, we obtain $\qquad P'C' = \frac{a'b'}{c'}$

and some equalities similar to equation (11.5) and (11.6). Then, we reduce:

$$\frac{c'^2 \, \triangle \mathbf{P'C'Q'}}{\triangle \mathbf{A'B'C'}} = a'^2 + b'^2 - c'^2 \qquad (11.8)$$

Comparing (11.7) with (11.8), we notice that $\angle PCQ = \angle P'C'Q'$, $\angle ACB$ and $\angle A'C'B'$ are supplementary and that

$$P'C' = \frac{a'b'}{c'}, PC = \frac{ab}{c}$$

By using the Co-Angle Theorem, we have:

$$\frac{a^2 + b^2 - c^2}{a'^2 + b'^2 + c'^2} = \frac{c^2}{c'^2} \bullet \frac{\triangle PCQ}{\triangle P'C'Q'} \bullet \frac{\triangle A'B'C'}{\triangle ABC}$$

$$= \frac{c^2}{c'^2} \bullet \frac{PC \bullet QC}{P'C' \bullet Q'C'} \bullet \frac{a'b'}{ab}$$

$$= \frac{c^2}{c'^2} \bullet \left(\frac{ba}{c}\right)^2 \bullet \left(\frac{c'}{a'b'}\right)^2 \bullet \frac{a'b'}{ab}$$

$$= \frac{a'b'}{ab} = \frac{\triangle ABC}{\triangle A'B'C'}$$

So far, we have already proved that when $\angle ACB$ and $\angle A'C'B'$ are supplementary,

$$\frac{a^2 + b^2 - c^2}{\triangle ABC} = -\frac{a'^2 + b'^2 - c'^2}{\triangle A'B'C'} \qquad (11.9)$$

Next we consider the case when $\angle ACB = \angle A'C'B'$.

We take any $\triangle XYZ$ such that

$\angle XYZ = 180° - \angle ACB$. Then, we have

$$\frac{a^2 + b^2 - c^2}{\triangle ABC} = -\frac{x^2 + y^2 - z^2}{\triangle XYZ} = \frac{a'^2 + b'^2 - c'^2}{\triangle A'B'C'} \qquad (11.10)$$

The proof of Pythagorean Difference Theorem has been completed.

108

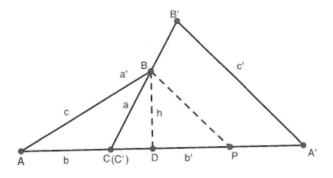

FIGURE 11.7

As an example, let's consider the special case where $\angle ACB = 90°$ in the proof above. Then, we have $\angle 1 + \angle 2 = \angle ACB = 90°$ in the first figure of Fig.11.6. So, $\angle PCQ = 0°$. The Pythagorean Theorem is obtained immediately from (11.6). Therefore, the Pythagorean Theorem is the special case of Pythagorean Difference Theorem.

If the Pythagorean Theorem is proved first, we can derive the Pythagorean Difference Theorem from the Pythagorean Theorem. Then, the proof is relatively simple. As shown in Fig.11.7, $\angle ACB$ and $\angle A'C'B'$ are supplementary. And let segments CB and $C'B'$ coincide. Draw a line through B parallel to $B'A'$ intersecting $A'C'$ at P. Then, there exists $k > 0$ such that

$$\frac{BC}{a'} = \frac{PC}{b'} = \frac{PB}{c'} = k$$

Draw the altitude BD of $\triangle ABC$ such that $BD = h$. By the Pythagorean Theorem, we have

$$\begin{cases} c^2 - (b + CD)^2 = h^2 = a^2 - CD^2 & (11.11) \\ BP^2 - (PC - CD)^2 = h^2 = a^2 - CD^2 & (11.12) \end{cases}$$

Reducing (8.11) and (8.12) respectively, we have

$$\begin{cases} c^2 - b^2 - 2b \cdot CD = a^2 & (11.13) \\ BP^2 - PC^2 + 2PC \cdot CD = a^2 & (11.14) \end{cases}$$

Substituting $BC = a = ka'$, $PC = kb'$, $PB = kc'$ into (11.14), we have

$$k^2 c'^2 - k^2 b'^2 + 2kb' \cdot CD = k^2 a'^2 \tag{11.15}$$

From (8.13) and (8.15) respectively, we have

$$\begin{cases} a^2 + b^2 - c^2 = -2b \cdot CD \\ a'^2 + b'^2 - c'^2 = \dfrac{2b' \cdot CD}{k} \end{cases} \tag{11.16} \tag{11.17}$$

Comparing two equations above, we have

$$\frac{a^2 + b^2 - c^2}{a'^2 + b'^2 - c'^2} = -\frac{kb}{b'} = -\frac{ab}{a'b'} = -\frac{\triangle\,ABC}{\triangle\,A'B'C'}$$

This is exactly what we want to prove.

ADDITIONAL PROBLEMS

[P11.1] Given the two sides a, b and the median l on side b in $\triangle$ ABC,

Determine: side c.

[P11.2] Prove: If

$$\frac{a^2 + b^2 - c^2}{a'^2 + b'^2 - c'^2} = \frac{\triangle\,ABC}{\triangle\,A'B'C'}$$

then $\angle C = \angle C'$.

[P11.3] Prove: When $\angle C \geq \angle C'$,

$$\frac{a'^2 + b'^2 - c'^2}{\triangle\,A'B'C'} \geq \frac{a^2 + b^2 - c^2}{\triangle\,ABC}$$

[P11.4] Prove:

$$\left(\frac{a^2 + b^2 - c^2}{2ab} \right)^2 + \left(\frac{2\,\triangle\,ABC}{ab} \right)^2 = 1$$

Then derive Heron's Formula from it.

CHAPTER 12

TRIANGLES AND CIRCLES

It would be a bit monotonous if there were only lines in geometric figures. Once circles appear, figures become more lively and colorful.

There are a plenty of properties of circles, the most important of which is the **Angle at the Circumference Theorem**: all angles whose vertex lies on the circumference are equal if they are subtended by the same arc or by an equal arc. In addition, they are equal to half of the angle of one whose vertex is at the center and is subtended by the same arc.

From the angle at the circumference theorem, we may immediately obtain a series of important properties of a circle: Alternate Segment Theorem, Interior Angle of the Circle Theorem, Exterior Angle of the Circle Theorem, Intersecting Chords Theorem, Tangent-Secant Theorem, Intersecting Secants Theorem……. You will learn them in geometry class.

Actually, in geometric figures, all of the circle's properties may be derived ultimately from the properties of the isosceles triangle. Think about it! Take two points on a circle, along with the circle's center, and you have an isosceles triangle. Obviously, the triangle is the more basic figure.

For example, there is an important theorem about the circle: the diameter perpendicular to a chord bisects this chord and the subtended arc. This can be interpreted as a theorem about the isosceles triangle: the altitude on the base of an isosceles triangle bisects the base and vertex angle.

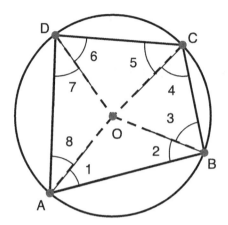

FIGURE 12.1

Also, the theorem, "the opposite angles of an inscribed quadrilateral of a circle are supplementary", is usually derived from the Angles at the Circumference Theorem, which uses the properties of circle. In fact, without the properties of circle, it is still simple to prove it directly using the properties of the isosceles triangle. As shown in Fig.12-1, to prove $\angle A + \angle C = 180°$, just notice that the sum of $\angle A$, $\angle B$, $\angle C$, $\angle D$ is 360°, and

$$\angle A + \angle C = \angle 1 + \angle 8 + \angle 4 + \angle 5$$

$$= \angle 2 + \angle 7 + \angle 3 + \angle 6$$

$$= \angle B + \angle D$$

We know immediately $\angle A + \angle C$ and $\angle B + \angle D$ are both equal to half of 360°! Here, we needn't refer to angles at the circumference.

If the center of circle O is in the exterior of the quadrilateral $ABCD$, as shown in Fig.12.2, the proof needs some modification, which is:

$$\angle A + \angle C = \angle 8 - \angle 1 + \angle 4 + \angle 5$$

$$= \angle 7 - \angle 2 + \angle 3 + \angle 6$$

$$= \angle 3 - \angle 2 + \angle 7 + \angle 6$$

$$= \angle B + \angle D$$

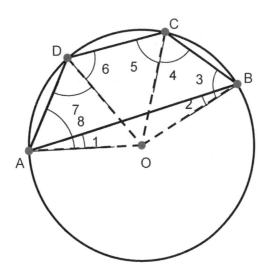

FIGURE 12.2

Thus, we may not only avoid the use of the Angle at the Circumference Theorem, but we also can deduce it. As shown in Fig.12.3, to prove $\angle P = \angle Q$, both of which subtend the same arc, it's sufficient to figure out that each angle is supplementary with $\angle C$.

In a word, the properties of the circle can be deduced from the properties of the isosceles triangle. However, a circle is a new kind of figure which can lead us to put forward new problems such as the calculation of its area and circumference, and offer new methods to us. If everything starts with the triangle, not only do things turn troublesome, but also the essence of things will be covered up in tedious reasoning. The statement, "The opposite angles of an inscribed quadrilateral of a circle are supplementary", is indeed intuitive and concise. It would be expressed as follows if we didn't involve the circle: "If the respective distance from a point to four vertex angles of a quadrilateral is the same, the opposite angles of the quadrilateral are supplementary." It's not beautiful at all.

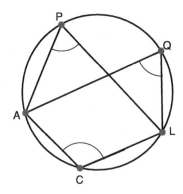

FIGURE 12.3

Thus, we need to not only know that the triangle is more basic, but also to attach importance to the role of the circle. By the angle at the circumference theorem, there are plenty of co-angle triangles in figures where circles are involved. To solve geometric problems about circles, it is crucial to flexibly apply the Co-Angle Theorem! Let's see a simple example, where we use the Co-Angle Theorem to deduce the Intersecting Chords Theorem, Tangent-Secant Theorem, and the Intersecting Secants Theorem.

[Ex12.1] Given: two lines through point P intersect the same circle at A, B and C, D respectively.

Prove: $PA \cdot PB = PC \cdot PD$.

Proof: Connecting AD and BC, then we have

$$\angle PAD = \angle PCB, \ \angle PDA = \angle PBC$$

Applying the Co-Angle Theorem to$\triangle PAD$ and $\triangle PBC$:

$$\frac{PA \cdot AD}{PC \cdot BC} = \frac{\triangle PAD}{\triangle PBC} = \frac{PD \cdot AD}{PB \cdot BC}$$

we have

$$\frac{PA}{PC} = \frac{PD}{PB}$$

that is, $PA \cdot PB = PC \cdot PD.$

This method is suitable for the three cases in Fig.12.4, where case (3) is the special case where point C coincides with point D. If point A coincides with point B, we also prove that "the lengths of two tangent line segments from a point to a circle are equal".

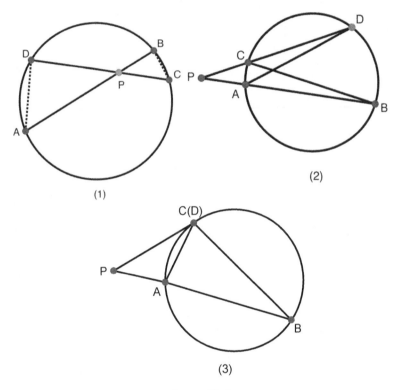

FIGURE 12.4

The results of the example above are probably familiar to readers, so we haven't put forward any new propositions. In the following, we will introduce a very useful proposition which may be unfamiliar to them.

[Ex12.2] Prove: the product of two sides of a triangle divided by the height of the third side is equal to the diameter of the circumcircle of the triangle.

Given: $\odot O$ is the circumcircle of $\triangle ABC$ and CD is the height of $\triangle ABC$ (as shown in Fig.12.5).

Prove: The diameter of $\odot O$ is $\quad d = \dfrac{AC \cdot BC}{CD}$

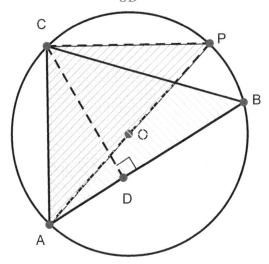

FIGURE 12.5

Proof: Draw a diameter of $\odot O$ through point A, denoted as AP. Then $AC \perp PC$ and $\angle APC = \angle ABC$. By Co-Angle Theorem, we have:

$$\frac{AC \cdot PC}{CD \cdot AB} = \frac{\triangle APC}{\triangle ABC} = \frac{AP \cdot PC}{BC \cdot AB}$$

$$\therefore \frac{AC}{CD} = \frac{AP}{BC}$$

$$\therefore d = AP = \frac{AC \cdot BC}{CD}$$

The following question is discussed explicitly in some books and journals. If we use the result of the example above, it will be very easy to solve.

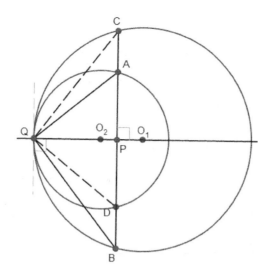

FIGURE 12.6

[Ex12.3] Given: $\odot\, O_2$ is internally tangent to $\odot\, O_1$ at Q. Take a point P on line $O_1 O_2$, and draw a perpendicular line to $O_1 O_2$ through P such that it intersects the two circles at A and B. The two circles have radii r_1 and r_2.

Determine the radius of the circumcircle of $\triangle\, QAB$.

Solution: Let points C and D be two more intersection points of line AB with $\odot\, O_1$ and O_2 (as shown in Fig.12.6). Then, we have

$$2r_1 = \frac{QB \cdot QC}{QP}, \qquad 2r_2 = \frac{QA \cdot QD}{QP}$$

Since AB is perpendicular to $O_1 O_2$, $QB = QC$, $QA = QD$. Let r be the radius of the circumcircle of $\triangle\, QAB$, and we have

$$2r = \frac{QA \cdot QB}{QP} = \sqrt{\frac{QA^2 \cdot QB^2}{QP^2}}$$

$$= \sqrt{\frac{QA \cdot QD}{QP} \cdot \frac{QB \cdot QC}{QP}}$$

$$= 2\sqrt{r_1 r_2}$$

$$\therefore r = \sqrt{r_1 r_2}$$

Here, it is unexpected that the radius of the circumcircle of $\triangle QAB$ is constant, independent of the position of P. Of course, PQ must be less than or equal to the diameter of $\odot O_2$. Otherwise, point A does not exist.

In following will introduce a proposition which is more useful for applications.

[Ex12.4] If the circumcircles of $\triangle ABC$ and $\triangle A'B'C'$ are the same or of equal radius, then

$$\frac{\triangle ABC}{\triangle A'B'C'} = \frac{BC \cdot CA \cdot AB}{B'C' \cdot C'A' \cdot A'B'}$$

Proof 1: Let the diameter of circumcircles of $\triangle ABC$ and $\triangle A'B'C'$ be d and d' respectively, the height on side AB of $\triangle ABC$ be h and the height on side $A'B'$ of $\triangle A'B'C'$ be h'. Then,

$$\frac{BC \cdot AC}{h} = d = d' = \frac{B'C' \cdot A'C'}{h'}$$

With

$$h = \frac{2 \triangle ABC}{AB}, \qquad h' = \frac{2 \triangle A'B'C'}{A'B'}$$

and substituting them into the equations above, we have

$$\frac{BC \cdot AC \cdot AB}{\triangle ABC} = \frac{B'C' \cdot A'C' \cdot A'B'}{\triangle A'B'C'}$$

This is what we want to prove.

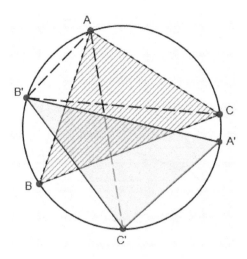

FIGURE 12.7

Proof 2: Let $\triangle ABC$ and $\triangle A'B'C'$ be inscribed in the same circle, as shown in Fig.12.7. By the Angle at the circumference Theorem and the Co-Angle Theorem, we have:

$$\frac{\triangle ABC}{\triangle A'B'C'} = \frac{\triangle ABC}{\triangle AB'C} \cdot \frac{\triangle AB'C}{\triangle AB'C'} \cdot \frac{\triangle AB'C'}{\triangle A'B'C'}$$

$$= \frac{AB \cdot BC}{AB' \cdot B'C} \cdot \frac{AC \cdot B'C}{AC' \cdot B'C'} \cdot \frac{AB' \cdot AC'}{A'B' \cdot A'C'}$$

$$= \frac{AB \cdot BC \cdot AC}{A'B' \cdot B'C' \cdot A'C'}$$

The advantage of proof 1 is that it does not require a figure. Yet, proof 2 is more direct, which only uses Angle at the Circumference Theorem and Co-Angle Theorem instead of the diameter formula of the circumcircle of a triangle deduced by Ex12.2. In the following, we call triangles with the same or equal circumcircle as **Co-Circle Triangles**. Ex.12.4 may be called the **Co-Circle Theorem**.

With the Co-Circle Theorem, we can deduce the diameter formula (which is left as an exercise). The following examples are very typical ones where the Co-Angle Theorem is used along with the angle at the circumference theorem.

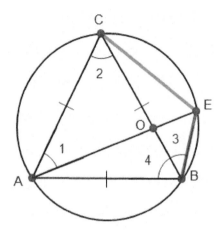

FIGURE **12.8**

[Ex12.5] Given: $\triangle ABC$ is an equilateral triangle. Take any point P on $\overset{\frown}{BC}$ of its circumcircle, and connect AP.

Prove: $AP = BP + CP$.

Proof: Suppose AP intersects BC at point O, as shown in Fig.12.8, and we have

$$\angle AOB = \angle 1 + \angle 2 = \angle 3 + \angle 4$$

$$= \angle APB = 180° - \angle ACP$$

Thus, we have

$$\frac{\triangle PAB + \triangle PAC}{\triangle AOB} = \frac{PB \cdot AB + PC \cdot AC}{AO \cdot BO}$$

$$= \frac{(PB + PC)AB}{AO \cdot BO} \qquad (12.1)$$

On the other hand:

$$\frac{\triangle PAB + \triangle PAC}{\triangle AOB} = \frac{PA}{AO} + \frac{\triangle PAC}{\triangle AOC} \cdot \frac{\triangle AOC}{\triangle AOB}$$

$$= \frac{PA}{AO} + \frac{PA}{AO} \cdot \frac{OC}{OB}$$

$$= \frac{PA \cdot BC}{AO \cdot BO} \qquad (12.2)$$

Comparing the right-hand side of (12.1) and (12.2), we have

$$PA = PB + PC$$

This example can also be proved smoothly by Co-Angle Theorem, for example:

$$1 = \frac{\triangle PCA + \triangle PBA}{\triangle OAB + \triangle OBP + \triangle OPC + \triangle OAC}$$

$$= \frac{\triangle PCA + \triangle PBA}{\triangle OAB} \cdot \frac{\triangle OAB}{\triangle OAB + \triangle OBP + \triangle OPC + \triangle OAC}$$

$$= \frac{(PB + PC)AB}{AO \cdot BO} \cdot \frac{AO \cdot BO}{AO \cdot BO + BO \cdot PO + PO \cdot CO + CO \cdot AO}$$

$$= \frac{(PB + PC) \cdot AB}{AP \cdot BC}$$

$$= \frac{PB + PC}{AP}$$

This example may prompt us to summarize a law:

If an intersecting angle of two diagonal lines of quadrilateral *ABCD* is equal or supplementary to an angle $\angle Q$ of $\triangle PQR$, we have

$$\frac{S_{ABCD}}{\triangle PQR} = \frac{AC \cdot BD}{PQ \cdot RQ}$$

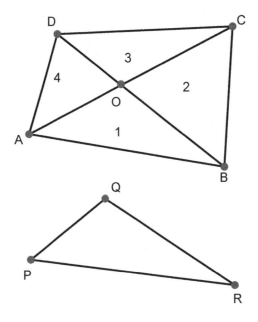

FIGURE 12.9

The proof is simple, since Ex12.5 has already suggested how to prove it. As shown in Fig.12.9:

$$\frac{S_{ABCD}}{\triangle PQR} = \frac{\triangle 1 + \triangle 2 + \triangle 3 + \triangle 4}{\triangle PQR}$$

$$= \frac{BO \cdot (AO + CO) + DO \cdot (AO + CO)}{PQ \cdot RQ}$$

$$= \frac{(AO + CO)(BO + DO)}{PQ \cdot RQ}$$

$$= \frac{AC \cdot BD}{PQ \cdot RQ}$$

With the law above, the proof of Ex12.5 may be more concise.

$$1 = \frac{\triangle PAB + \triangle PAC}{S_{ABCD}} = \frac{AB \cdot PB + AC \cdot PC}{AP \cdot BC}$$

$$= \frac{PB + PC}{AP}$$

122

Such a brief expression grasps the essence of the problem. We will often find some little tricks in the process of solving problems. Summarizing these little tricks in time and refine them into laws, will help us to accumulate experience and then form skills and patterns. Actually, the Co-Side Theorem, Co-Angle Theorem and Pythagorean Difference Theorem, etc. can all be summarized likewise.

[Ex12.6] (*Ptolemy's Theorem*) **Given:** convex quadrilateral ABCD is inscribed in circle O.

Prove: $AB \cdot CD + AD \cdot BC = AC \cdot BD$.

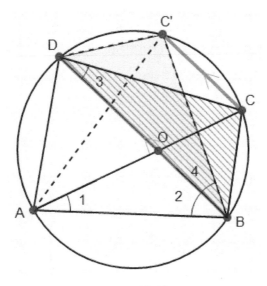

FIGURE **12.10**

Proof: As shown in Fig.12.10, draw a line parallel to *BD* from point *C* that it intersects the circle at point C'. Obviously, we have

$$DC' = BC$$

$$BC' = CD$$

$$\triangle BDC' = \triangle BDC$$

Denote the intersection point of *AC* and *BD as O*. Then, we have

$$\angle AOD = \angle 1 + \angle 2 = \angle 3 + \angle 2 = \angle 4 + \angle 2$$

$$= \angle ABC' = 180^\circ - \angle ADC'$$

$$\therefore 1 = \frac{\triangle ADC' + \triangle ABC'}{S_{ABCD}}$$

$$= \frac{AD \cdot DC' + AB \cdot BC'}{AC \cdot BD}$$

$$= \frac{AD \cdot BC + AB \cdot CD}{AC \cdot BD}$$

The idea of this proof is: when noticing the right-hand side of the proved equation is the product of two diagonal lines, you should then think of the area of quadrilateral *ABCD*. However, since any term in the left-hand side is the product of two segments, which are opposite sides of quadrilateral *ABCD*, it is difficult to determine the area. Can we get the two segments together? Draw a parallel line through point *C* such that the position of *BC* and *DC* are exchanged, and the problem is thus solved.

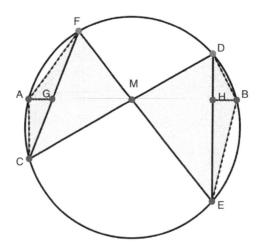

FIGURE 12.11

[Ex12.7] (**Butterfly Theorem**) **Given:** as shown in Fig.12.11, *M* is the midpoint of chord *AB*. Draw another two chords *CD* and *EF* through *M* and connect *C* to *F* and *D* to *E* intersecting *AB* at points *G and H* respectively.

Prove: $MG = MH$.

Proof: It is sufficient to prove

$$\frac{MG}{AG} = \frac{MH}{BH}$$

First using Co-Side Theorem and then using Co-Angle Theorem along with Co-Circle Theorem, we have

$$\frac{MG}{AG} \cdot \frac{BH}{MH} = \frac{\triangle MCF}{\triangle ACF} \cdot \frac{\triangle BDE}{\triangle MDE}$$

$$= \frac{\triangle MCF}{\triangle MDE} \cdot \frac{\triangle BDE}{\triangle ACF}$$

$$= \frac{MC \cdot MF}{MD \cdot ME} \cdot \frac{BD \cdot DE \cdot BE}{AC \cdot CF \cdot AF}$$

$$= \frac{MC}{MD} \cdot \frac{MF}{ME} \cdot \frac{MD}{MA} \cdot \frac{ME}{MC} \cdot \frac{MB}{MF}$$

$$= \frac{MB}{MA} = 1$$

The Butterfly Theorem is a well-known problem. The proof above is novel and concise. The idea goes as follows: eliminate G, H first, and then transform all chords into the ratios of segments with the endpoint M, which makes the answer clear.

ADDITIONAL PROBLEMS

[P12.1] Suppose d is the diameter of the circumcircle of $\triangle ABC$.

Prove:
$$d = \frac{BC \cdot AC \cdot AB}{2 \triangle ABC}$$

[P12.2] As shown in Fig.12.4, draw two lines through point P such that they intersect a circle at A,B,C,D respectively.

Prove:
$$\frac{AC \cdot AD}{BC \cdot BD} = \frac{PA}{PB}$$

[P12.3] Given: the angles included by the two diagonal lines of quadrilateral $ABCD$ are equal to the corresponding angles included by the two diagonal lines of quadrilateral $WXYZ$.

Prove:

$$\frac{S_{ABCD}}{S_{WXYZ}} = \frac{AC \cdot BD}{WY \cdot XZ}$$

[P12.4] In Ex12.7, if M is not the midpoint of AB.

Prove:

$$\frac{MG}{MH} \cdot \frac{MA}{MB} = \frac{AG}{BH}$$

[P12.5] In Ex12.7, if the intersecting point of DC and EF is not on chord AB, what happens to the conclusion of the problem? (Refer to exercise 7.8.)

CHAPTER 13

TRIANGLES AND CIRCLES (CONTINUED)

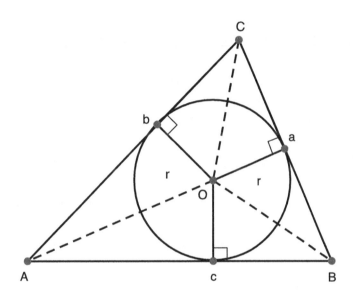

FIGURE **13.1**

If a circle is inscribed in a triangle, there are simple relationships between triangle area $\triangle$, three sides a, b, c and the radius r. As shown in Fig.13.1, we have

$$\triangle ABC = \triangle OAB + \triangle OBC + \triangle OAC$$

$$= \frac{r}{2}(a + b + c)$$

$$\therefore \quad r = \frac{2\triangle}{a + b + c}$$

Using Ex12.2 in the previous section, we can determine the radius R of the circumcircle of $\triangle ABC$, which is

$$R = \frac{abc}{4\,\triangle}$$

These relationships are useful for solving problems.

[Ex13.1] Prove: the radius of the circumcircle of a triangle is greater than or equal to twice the radius of its incircle.

Proof: Applying Heron's Formula (refer to Ex11.1), we have

$$
\begin{aligned}
\frac{R}{r} &= \frac{abc(a + b + c)}{8\,\triangle^2} \\
&= \frac{2abc(a + b + c)}{4b^2c^2 - (b^2 + c^2 - a^2)^2} \\
&= \frac{2abc(a + b + c)}{(2bc + b^2 + c^2 - a^2)(2bc - b^2 - c^2 + a^2)} \\
&= \frac{2abc(a + b + c)}{(a + b + c)(a + b - c)(a - b + c)(b - a + c)} \\
&= 2 \cdot \frac{a}{\sqrt{a^2 - (b - c)^2}} \cdot \frac{b}{\sqrt{b^2 - (a - c)^2}} \cdot \frac{c}{\sqrt{c^2 - (a - b)^2}} \geq 2
\end{aligned}
$$

The first part of the proof is not surprising. However, the last step is very smart, which compares the value of abc with that of
$$(a + b - c)(a - b + c)(b - a + c)$$
When comparing, one skillfully decomposes each factor in the denominator into two terms and then rearranges them, the process of which can be written explicitly as follows:

$$(a + b - c)(a - b + c)(b - a + c)$$
$$= \sqrt{(a + b - c)^2(a - b + c)^2(b - a + c)^2}$$

$$= \sqrt{(a + b - c)(a - b + c)} \cdot \sqrt{(b + a - c)(b - a + c)}$$

$$\cdot \sqrt{(c+a-b)(c-a+b)}$$
$$= \sqrt{a^2 - (b-c)^2} \cdot \sqrt{b^2 - (a-c)^2} \cdot \sqrt{c^2 - (a-b)^2}$$

Algebraic transformations can be very useful in solving geometry problems. In this problem, we may alternatively avoid algebraic computation and use the relationship of the area and perimeter of a triangle. It is sufficient to prove that for any triangle with a given perimeter, the area of an equilateral triangle is the biggest. Then look at the radius of an incircle of an equilateral triangle. This question is left as an exercise.

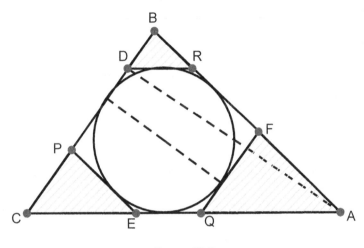

FIGURE 13.2

[Ex13.2] (International Olympics Mathematics Competition, 1964) As shown in Fig.13.2, a circle is inscribed in $\triangle ABC$, three sides of which are a, b and c respectively. Draw three tangent lines to the circle, each parallel to the opposite side. Three small triangles are carved out in $\triangle ABC$. Determine the sum of the area of the incircles of the four triangles above.

Solution: As shown in Fig.13.2, suppose

$$s = \frac{1}{2}(a+b+c)$$

Then the radius of the incircle of $\triangle ABC$ is solved for:

$$r = \frac{\triangle ABC}{s} \cdot \frac{ah_a}{2s},$$

where h_a is the height on side a of $\triangle ABC$ Thus, the ratio of r_A, the radius of the incircle of $\triangle AQF$, to r is as follows:

$$\frac{r_A}{r} = \frac{h_a - 2r}{h_a} = 1 - \frac{a}{s}$$

Hence, the sum of the area of 4 circles follows:

$$\pi(r^2 + r_A{}^2 + r_B{}^2 + r_C{}^2)$$
$$= \pi\left[1 + \left(1 - \frac{a}{s}\right)^2 + \left(1 - \frac{b}{s}\right)^2 + \left(1 - \frac{c}{s}\right)^2\right]r^2$$
$$= \frac{\pi}{s^3}(s - a)(s - b)(s - c)(a^2 + b^2 + c^2)$$

Simplification of the last step in the equation above is due to expanding

$$\left(1 - \frac{a}{s}\right)^2, \left(1 - \frac{b}{s}\right)^2, \left(1 - \frac{c}{s}\right)^2$$

and using	$r = \dfrac{\triangle ABC}{s}$	along with Heron's Formula.

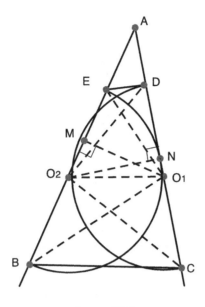

FIGURE 13.3

[Ex13.3] As shown in Fig.13.3, in $\triangle ABC$, take points O_1 and O_2 on side AC and AB respectively. Draw a circle with center O_1 such that it is tangent to AB and intersects AC at D and C. Draw a circle with center O_2 such that it is tangent to AC and it intersects AB at B and E.

Prove: $BC \parallel DE$.

Proof: Suppose the tangent point of $\odot O_2$ to AC is N, and the tangent point of $\odot O_1$ to AD is M. Then, we have

$$\triangle DO_1O_2 = \frac{1}{2}O_2N \cdot DO_1 = \frac{1}{2}O_2E \cdot O_1M$$
$$= \triangle EO_1O_2$$
$$\therefore \quad DE \parallel O_1O_2$$

Since O_1, O_2 are the midpoints of BE and CD respectively, we have

$$\triangle BO_1O_2 = \triangle EO_1O_2 = \triangle DO_1O_2 = \triangle CO_1O_2$$
$$\therefore \quad BC \parallel O_1O_2$$
$$\therefore \quad BC \parallel DE$$

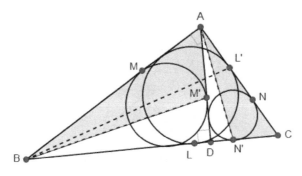

FIGURE 13.4

[Ex13.4] In $\triangle ABC$, $\angle A = 90°$, and AD is the altitude on side BC. The incircles of three right triangles $\triangle ABC$, $\triangle ABD$, $\triangle ACD$ are tangent to their hypotenuses at L, M, N, respectively (as shown in Fig.13.4).

Prove:

$$AM \cdot AB + CN \cdot AC = CL \cdot BC$$

Such a question can certainly be calculated directly. Given AB, AC, by Pythagorean Theorem we can calculate BC. With $AD \times BC = AB \times AC$, we can calculate AD. Using the relevant proportion, we can also calculate BD and CD. Since the three sides of every triangle are known, the length of tangent lines AM, CN, CL can be calculated. The question can be solved by substituting the values into the equation. But since it's very troublesome to calculate, we should seek some shortcuts.

Provided here are two ideas. Please finish the proof by yourself.

Idea One: With $\triangle ABC \sim \triangle DBA \sim \triangle DAC$, we may prove:

$$\frac{AM}{AB} = \frac{CN}{AC} = \frac{CL}{BC} = k$$

So $AM = kAB$, $CN = kAC$, $CL = kBC$. Then use the Pythagorean Theorem to complete the proof.

Idea Two: Transform the product of line segments into the areas. Suppose that the tangent point on side AD of $\triangle ABD$ is M', the tangent point on side CD of $\triangle ADC$ is N', and the tangent point on side AC of $\triangle ABC$ is L'. Then, by the Co-Angle Theorem we have (Fig.13.4)

$$\frac{\triangle\,ABM' + \triangle\,ACN'}{\triangle\,BCL'} = \frac{AM \cdot AB + CN \cdot AC}{CL \cdot BC}$$

Then, it is sufficient to prove $\triangle\,ABM' + \triangle\,ACN' = \triangle\,BCL'$. We can achieve this purpose by proving the following equation.

$$\frac{\triangle\,ABM'}{\triangle\,ABD} = \frac{\triangle\,ACN'}{\triangle\,ACD} = \frac{\triangle\,BCL'}{\triangle\,ABC}$$

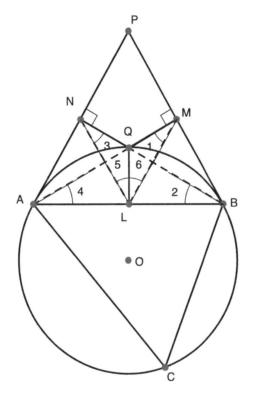

FIGURE 13.5

[Ex13.5] Suppose $\odot\,O$ is the circumcircle of an acute triangle $\triangle\,ABC$. AB is one side of $\triangle\,ABC$. Draw tangent lines to $\odot\,O$ through points A and B such that they intersect at

point P. Take any point Q on $\overset{\frown}{AB}$ and draw three perpendicular lines through Q to three sides of $\triangle\,PAB$, the feet of which are L, M, N, respectively. (as shown in Fig.13.5)

Prove: $LQ^2 = MQ \cdot NQ$.

Proof: by the Angle at the Circumference Theorem and Alternate Segment Theorem, we have:

$$\angle PAB = \angle ACB = \angle PBA$$
$$\angle AQB = 180° - \angle C$$

And since $\angle QLB = \angle QMB = \angle QLA = \angle QNA = 90°$, we have

$$\angle LQM = 180° - \angle PBA$$
$$\angle LQN = 180° - \angle PAB$$
$$\angle 1 = \angle 2, \angle 3 = \angle 4$$
$$\angle LQN = \angle LQM = \angle AQB$$
$$\therefore \angle 5 = \angle 2 = \angle 1, \angle 6 = \angle 4 = \angle 3$$

By the Co-Angle Theorem, we have

$$\frac{MQ \cdot ML}{LQ \cdot LN} = \frac{\triangle MQL}{\triangle LQN} = \frac{LQ \cdot ML}{NQ \cdot LN}$$
$$\therefore \frac{MQ}{LQ} = \frac{LQ}{NQ}$$

that is,

$$LQ^2 = MQ \cdot NQ$$

Ex13.5 adequately exemplifies this approach: we should first understand the relationship of equality and supplementarity among angles, and then try to apply the Co-Angle Theorem.

The following question is of a special style.

As is known, the line segment joining the midpoints of two sides of a triangle—the midsegment— is equal to half of the third side. And in the triangle made up of the three midsegments, each side of it is half of the corresponding side of the original triangle. How about the converse?

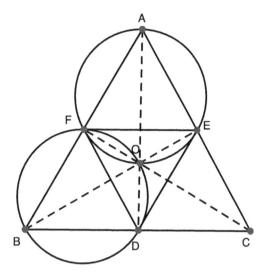

FIGURE 13.6

[Ex13.6] In $\triangle ABC$, points D, E and F are on side BC, CA, AB respectively such that $DE = \frac{1}{2}AB$, $EF = \frac{1}{2}BC$, $DF = \frac{1}{2}AC$ (as shown in Fig.13.6).

Prove: D, E and F are the midpoints of sides BC, CA, AB, respectively.

Proof: Take a point O such that $\angle DOF$ and $\angle B$ are supplementary and likewise for $\angle FOE$ and $\angle A$. It can be easily achieved if we take point O as an intersection point of the circumcircles of $\triangle BDF$ and $\triangle AFE$.

After calculation, we know that $\angle DOE$ and $\angle C$ are supplementary, which means that point O is on the circumcircle of $\triangle DCE$.

In the following, we will prove that O is the circumcenter of $\triangle DCE$, and that $OF \perp AB$, $OD \perp BC$ and $OE \perp AC$.

Suppose r_A, r_B, r_C and R are the radii of the circumcircles of $\triangle AFE$, $\triangle BDF$, $\triangle CDE$ and $\triangle ABC$, respectively. By the Radius Formula of Circumcircles and the Co-Angle Theorem, we have:

$$\frac{R}{r_A} = \frac{AB \cdot AC \cdot BC}{AF \cdot AE \cdot FE} \cdot \frac{\triangle AFE}{\triangle ABC}$$
$$= \frac{AB \cdot AC \cdot BC}{AF \cdot AE \cdot FE} \cdot \frac{AF \cdot AE}{AB \cdot AC} = \frac{BC}{EF} = 2$$

So $r_A = \frac{1}{2}R$ Likewise, $r_B = r_C = \frac{1}{2}R$. From the properties of Co-Circle Triangles, we have:

$$\frac{AF}{BF} = \frac{\triangle AFO}{\triangle BFO} = \frac{AF \cdot FO \cdot AO}{BF \cdot FO \cdot BO} = \frac{AF \cdot AO}{BF \cdot BO}$$
$$\therefore AO = BO$$

Likewise, $BO = CO$. So, AO, BO and CO are the radii of the circumcircle of $\triangle ABC$, that is, the diameters of the circumcircles of $\triangle AFE$, $\triangle BDF$, $\triangle CDE$. Then, $\angle AFO = 90°$. Hence $AF = BF$. Likewise, we can prove D and E are the midpoints of BC and CA respectively.

This question shows that circles can help us to find shortcuts to solving problems. Although the question itself does not involve a circle, its essence is obvious when the appropriate circle is added. What makes us think of adding the circle?

Suppose the conclusion we proved is correct, and then D, E and F are the midpoints of the three sides. Draw perpendicular lines to BC, CA, AB at D, E and F respectively, the three should intersect at the circumcenter of $\triangle ABC$. But how can we find point O if we are not given that D, E, F are the midpoints of the three sides? It also requires thinking about other properties of point O. After observation, we find that $\angle FOE$ and $\angle A$ should be supplementary and that $\angle DOF$ and $\angle B$ should also be so. In order to find a point with such properties, we are led to add the respective circumcircle of $\triangle AEF$ and $\triangle BDF$.

When solving problems, such reflection on methodology is very useful.

ADDITIONAL PROBLEMS

[P13.1] Given: $\triangle PAB$ and $\triangle QAB$ are two inscribed triangles of the same circle. If P and Q are on the same side of AB and

$$|\angle PAB - \angle PBA| > |\angle QAB - \angle QBA|$$

Prove:
$$\frac{PA + PB + AB}{QA + QB + AB} > \frac{\triangle PAB}{\triangle QAB}$$

[P13.2] Prove by the conclusion above: the radius of the circumcircle of a triangle is greater than or equal to twice the radius of its incircle. In addition, the equality holds if and only if the triangle is an equilateral triangle.

[P13.3] As shown in Fig.13.3 - **Given:** $BC \parallel DE$ and the circle with diameter BE is tangent to AC.

Prove: the circle with diameter DC is tangent to AB.

[P13.4] Take any point P on the incircle of equilateral triangle $\triangle ABC$. Draw perpendicular line segments PX, PY, PZ through P to the three sides of $\triangle ABC$, where PX is the longest one.

Prove: $$\sqrt{PX} = \sqrt{PY} + \sqrt{PZ}.$$ (Hint: make use of Ex13.5.)

[P13.5] In the question above, if P is on an excircle of $\triangle ABC$, what conclusions can we draw?

CHAPTER 14

SUMMARY

Summarizing the tools of solving problems we have introduced, there are:

(1) **Co-Side Theorem -** If line PQ intersects AB at M, then

$$\frac{\triangle PAB}{\triangle QAB} = \frac{PM}{QM}$$

(2) **The Relationship of Parallel and Area -** If PQ ∥ AB, then $\triangle PAB = \triangle QAB$; Conversely, if $\triangle PAB = \triangle QAB$ and P and Q are on the same side of AB, then   PQ ∥ AB.

(3) **Point of Division Formula -** Suppose that point T is on line PQ and PT = λPQ. Then, for any two points A, B, if line segment PQ doesn't intersect line AB, we have:

$$\triangle TAB = \lambda \triangle QAB + (1 - \lambda) \triangle PAB$$

if line segment PQ intersects AB at M and T is on line segment PM, we have:

$$\triangle TAB = (1 - \lambda) \triangle PAB - \lambda \triangle QAB$$

(4) **Co-Angle Theorem -** If $\angle ABC = \angle A'B'C'$ or $\angle ABC + \angle A'B'C' = 180°$, then we have

$$\frac{\triangle ABC}{\triangle A'B'C'} = \frac{AB \cdot BC}{A'B' \cdot B'C'}$$

(5) **Co-Angle Inequality -** If $\angle ABC > \angle A'B'C'$ and the sum of two angles is less than 180°, then

$$\frac{\triangle ABC}{\triangle A'B'C'} > \frac{AB \cdot BC}{A'B' \cdot B'C'}$$

(6) *Co-Angle Converse Theorem -*

If $\dfrac{\triangle ABC}{\triangle A'B'C'} = \dfrac{AB \cdot BC}{A'B' \cdot B'C'}$ then $\angle ABC$ and $\angle A'B'C'$ are equal or supplementary.

(7) *Pythagorean Difference Theorem* - If $\angle ACB = \angle A'C'B'$ or they are

supplementary then $\dfrac{a^2 + b^2 - c^2}{\triangle ABC} = \pm \dfrac{a'^2 + b'^2 - c'^2}{\triangle A'B'C'}$

If two angles are equal, then we take the positive sign; if two angles are supplementary, then we take the negative sign, where a, b, c and $a', b' c'$ denote sides BC, CA, AB and $B'C', C'A', A'B'$ respectively.

(8) *Formula for the Diameter of the Circumcircle of a Triangle* - Suppose CD is the altitude on side AB of $\triangle ABC$, and then the diameter of the circumcircle of $\triangle ABC$

is $\qquad d = \dfrac{AC \cdot BC}{CD}$

(9) *Co-Circle Theorem* - If the inscribed circles of $\triangle ABC$ and $\triangle A'B'C'$ are the same or

equal then $\dfrac{\triangle ABC}{\triangle A'B'C'} = \dfrac{AB \cdot BC \cdot CA}{A'B' \cdot B'C' \cdot C'A'}$

In addition to these, do not forget to use Area Equations.

If you are familiar with these tools along with the fundamentals in geometry curriculums, especially the properties of parallel lines and the sum of triangle interior angles, Angle at the Circumference Theorem and so on, then most general geometric problems can be solved.

Although geometric problems involve proof, construction, calculation and so on, these methods have a lot in common. In order to show the reasonableness of construction problems, proofs are needed. If the results of calculation problems are given first, they become proofs. If you don't know the conclusion of proof problems, then they can be turned into calculation problems.

Proofs can be divided into two categories: equalities and inequalities. If you're to prove that two lines are parallel or perpendicular to each other, three points are collinear and four points lie on a circle, then these conclusions can be expressed by equalities. So, all of these are equality type problems. If the equal sign is explicitly placed in the conclusion, such as to prove that two angles or two line segments are equal and proportional expressions, then they are absolutely equality type problems. If you're to prove that a point is not on a line segment, in the interior or exterior of a circle, or that three line segments can constitute a triangle or a certain inequality, then these are inequality type problems.

Professor Wenjun Wu, a famous Chinese mathematician, founded the field of machine proofs of geometry theorems, creating a technique which is internationally called Wu's method. Equality type geometry theorems can be proved by computer using Wu's method. If you calculate by hand, the results can also be figured out as long as you calculate step-by-step patiently and meticulously.

For geometry inequalities, there hasn't been any effective machine proof. Usually, some skill is required. Because of this, geometry problems in the international mathematics competitions are often of the inequality type. The problem designers want candidates not to calculate rigidly but to come up with some smart ideas.

This book has mainly introduced the area method of solving geometry problems, which combines geometry and algebra. You should do the algebraic deduction and calculation while you make reference to the figures and notice the geometric quantities within. The advantages of this approach are that it not only looks straightforward but also that it has an almost fixed recipe with few auxiliary lines.

To convert geometric problems into algebraic problems, trigonometric functions are also a powerful tool. Here, although we haven't mentioned trigonometric functions, the area methods have a close relationship with trigonometric functions.

The Co-Angle Theorem tells us that if $\angle ABC$ and $\angle A'B'C'$ are equal or supplementary, then

$$\frac{\triangle ABC}{\triangle A'B'C'} = \frac{AB \cdot BC}{A'B' \cdot B'C'} \qquad \text{and} \qquad \frac{\triangle ABC}{AB \cdot BC} = \frac{\triangle A'B'C'}{A'B' \cdot B'C'}$$

This equality tells us that the ratio of the triangle area and the product of two sides depends only on the size of the included angle between those sides. If you know the concept of function, you will immediately know that this ratio is a function of the angle. Let's call it "the area coefficient" of the angle.

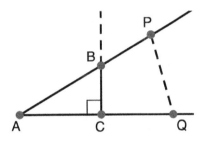

FIGURE 14.1

Suppose we are given an angle $\angle A$,　how can we find the area coefficient of $\angle A$? By the Co-Angle Theorem, take points P and Q on each side of $\angle A$, and it is sufficient to calculate

$$\frac{\triangle APQ}{AP \cdot AQ}$$

However, the area formula of any triangle is relatively complex. Since P and Q can be placed randomly, why must we create so much trouble for ourselves? We can manage to calculate the area of $\triangle APQ$ more conveniently. For what kind of triangle is the area easier to calculate? Of course, for the right triangle! As shown in Fig.14.1, if we take $\triangle ABC$ such that $\angle BCA = 90°$ instead of $\triangle APQ$, we have

$$\frac{\triangle ABC}{AB \cdot AC} = \frac{\frac{1}{2}AC \cdot BC}{AB \cdot AC} = \frac{1}{2} \cdot \frac{BC}{AB}$$

which shows us that the area coefficient of $\angle A$ is equal to half of the ratio of the opposite side to the hypotenuse.

If you have learned the fundamentals of trigonometric functions and sine, you know that the sine of $\angle A$ in a right triangle where $\angle A$ is an acute angle ($sinA$), is the ratio of the opposite side to the hypotenuse. If $\angle A$ is a right angle, then $sinA=1$. If two angles are supplementary, then their sines are equal. By this analysis, we draw the conclusion that the area coefficient of $\angle A$ is half of $sinA$. When you solve a problem by the Co-Angle Theorem, it is equivalent to using the sine function to solve the problem! Conversely, when you use trigonometric methods to solve a problem, the Co-Angle Theorem can be applied in place of using the sine.

So, what is the relationship between other trigonometric functions and the area, such as cosine and tangent? The three sides of $\triangle ABC$ are a, b, c. With these three sides, we can formulate several kinds of figures of different areas:

Each side can generate a square, the area of which is a^2, b^2 and c^2 respectively;

Every two sides can generate a rectangle, the area of which is bc, ca and ab respectively;

Three sides can formulate a triangle, the area of which is $\triangle ABC$.

Using these areas, we can make many inferences. We introduce the "Pythagorean Difference":

The Pythagorean Difference of $\angle ABC$ is $a^2 + c^2 - b^2$;

The Pythagorean Difference of $\angle BAC$ is $b^2 + c^2 - a^2$;

The Pythagorean Difference of $\angle ACB$ is $a^2 + b^2 - c^2$.

And we have proved the ratio $\dfrac{\triangle ABC}{a^2 + c^2 - b^2}$ depends only on the size of $\angle ABC$

and not on the length of a, b, c. So we can further prove the ratio, $\dfrac{a^2 + c^2 - b^2}{ac}$

is also the function of $\angle B$. Please prove that this ratio is twice that of $cosB$. The ratio,

$\dfrac{\triangle ABC}{a^2 + c^2 - b^2}$ is a quarter of $tanB$. Thus, there is a close relationship between area

and trigonometric functions. Nevertheless, this book will not further discuss problems with trigonometric functions.

ADDITIONAL PROBLEMS

[P14.1] In the **Point of Division Formula**, when T is on PQ extended, how would you calculate $\triangle TAB$? How about when PQ intersects AB at M and T is on MQ?

[P14.2] Prove by the area relationship: when $\alpha + \beta < 180°$ and $0 \le \alpha \le \beta$, we have $sin\alpha \le sin\beta$.

[P14.3] Prove by the Pythagorean Theorem: in any triangle $\triangle ABC$, we have

$$cosA = \frac{b^2 + c^2 - a^2}{2bc}$$

[P14.4] Prove by the area relationship: when $0 \le \alpha \le \beta < 180°$,
$$cos\alpha \ge cos\beta$$

CHAPTER 15

SELECTED AREA PROBLEMS FROM MATHEMATICS COMPETITIONS

So-called area problems are those that refer to areas. Geometry problems related to area are very common. The following area problems are only a small sample.

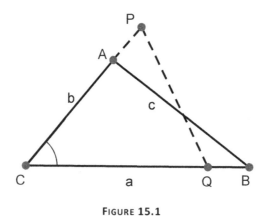

FIGURE 15.1

[Ex15.1] (Hungary Olympic Mathematics Competition, 1902)

Given: the area S and the vertex angle C of a triangle, what condition on the included sides a and b make the opposite side c shortest?

Solution: Draw an isosceles triangle whose area is S, and vertex angle is C, as shown in Fig.15.1. Assuming $\triangle ABC = \triangle PCQ$, let

$$c_0 = PQ, k = PC = QC$$

By Co-Angle Theorem, we have

$$1 = \frac{\triangle ABC}{\triangle PCQ} = \frac{ab}{k^2}$$

So ab $= k^2$. Applying Pythagorean Difference Theorem to $\triangle ABC$ and $\triangle PCQ$, we have

$$1 = \frac{\triangle ABC}{\triangle PCQ} = \frac{a^2 + b^2 - c^2}{k^2 + k^2 - c_0^2} = \frac{a^2 + b^2 - c^2}{2ab - c_0^2}$$

$$\therefore c^2 = c_0^2 + (a - b)^2 \, ,$$

which shows that when $a = b$, c is the smallest.

The key to solving this problem is to compare the $\triangle ABC$ with the isosceles triangle with the same area. It is not difficult to come up with. Intuitively, if the area is fixed and one included side is very short, then the other must be very long. So, the opposite side c is very long. To make c a minimum, a and b should not be such that one is long while the other is short. Thus, we've found the isosceles triangle! If you cannot come up with the idea of the isosceles triangle, you can compare $\triangle ABC$ with any triangle with an angle of $\angle C$. But you have to deduce with two more steps.

The proof is presented as follows:

Suppose P, Q are any two points on the two sides of $\angle C$. Let

$$PQ = h, PC = q, QC = p$$

Assume

$$\lambda = \frac{\triangle ABC}{\triangle PCQ} = \frac{ab}{pq}$$

Then ab $= \lambda pq$. By the Pythagorean Difference Theorem again, we have

$$\lambda = \frac{\triangle ABC}{\triangle PCQ} = \frac{a^2 + b^2 - c^2}{p^2 + q^2 - h^2}$$

$$\therefore c^2 = \lambda h^2 - \lambda(p^2 + q^2) + a^2 + b^2$$

$$= \lambda h^2 - \lambda(p^2 + q^2) + 2\,\lambda pq - 2ab + a^2 + b^2$$

$$= \lambda[h^2 - (p - q)^2] + (a - b)^2$$

Since λ, h, p, q are fixed, when $a = b$, c is the smallest.

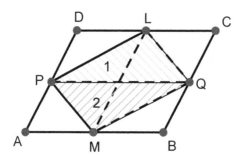

FIGURE 15.2

[Ex15.2] (Hungary Mathematics Competition, 1915)

Prove: it is impossible that the area of a triangle inscribed in a parallelogram is bigger than half of the area of the parallelogram.

This problem is easy. Let LMP be the inscribed triangle. As shown in Fig.15.2, draw a line parallel to *AB* through *P*, which intersects *BC* at *Q*. Line *PQ* divides $\triangle LMP$ into $\triangle 1$ and $\triangle 2$, where

$$\triangle 1 \leq \triangle PQL = \frac{1}{2}\,\square PQCD$$

$$\triangle 2 \leq \triangle PQM = \frac{1}{2}\,\square PQBA$$

$$\therefore \triangle 1 + \triangle 2 \leq \frac{1}{2}\left(\square PQCD + \square PQBA\right) = \frac{1}{2}\,\square ABCD$$

The proof is finished.

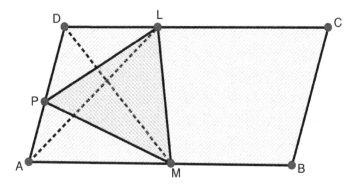

FIGURE **15.3**

Although the proof above is simple and beautiful, it is not as profound and general as the following. As shown in Fig.15.3, compare the areas of $\triangle AML$, $\triangle PML$, $\triangle DML$. If $AD \parallel ML$, then the areas of the three triangles are equal. If AD is extended such that it intersects ML, then

$$\triangle AML > \triangle PML.$$

if DA is extended such that it intersects LM, then

$$\triangle DML > \triangle PML$$

In all, one of $\triangle AML$ and $\triangle DML$ must be not less than $\triangle PML$. Suppose $\triangle AML \geq \triangle PML$, and then

$$\triangle PML \leq \triangle AML \leq \triangle ABL = \frac{1}{2}\square ABCD$$

Similarly, the question is solved. But the latter leads to a general idea:

Proposition 15.1: Suppose one vertex A of $\triangle ABC$ is in the interior or on the circumference of a polygon $P_1 P_2 \ldots P_n$, then there must exist certain
$$P_k(1 \leq k \leq n)$$
such that

$$\triangle P_k BC \geq \triangle ABC$$

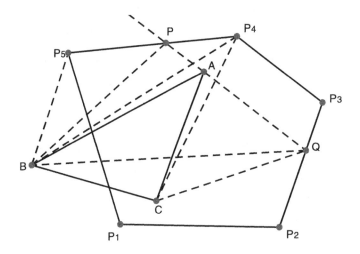

FIGURE 15.4

Proof: As shown in Fig.15.4, draw any line through point A, which intersects the circumference of polygon $P_1 P_2 \dots P_n$ at points P and Q. Then, the bigger of $\triangle PBC$ and $\triangle QBC$ is greater than or equal to $\triangle ABC$. Suppose $\triangle PBC \geq \triangle ABC$, and that point P is on side $P_l P_{l+1}$. Then, one of $\triangle P_l BC$ and $\triangle P_{l+1} BC$ is greater than or equal to $\triangle PBC$. So, it is greater than or equal to $\triangle ABC$.

Proposition 15.1 also has a simpler proof. As shown in Fig.15.5, draw L parallel to BC through A and take any vertex P_k such that it's not on the same side of l as B, C or it falls on L. Then, we have

$$\triangle P_k BC \geq \triangle ABC$$

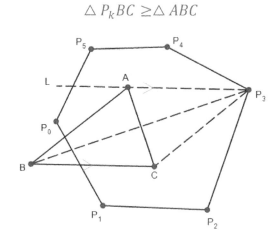

FIGURE 15.5

148

In the Fig.15.5, it's fine to take any one point of P_3, P_4, P_5.

Ex15.2 is very easy with the proposition 15.1 since we may use the vertices of the parallelogram to replace the vertices of $\triangle PML$ one by one. Whereas, the area of the triangle does not decrease! In fact, proposition 15.1 can help us to solve more difficult problems, for example the following.

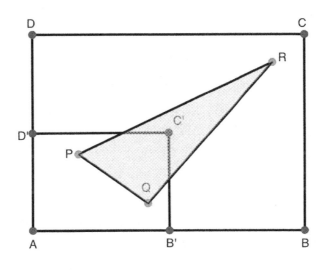

FIGURE 15. 6

[Ex15.3] (Mathematics Competition of Anhui Province, China, 1979) There are two rectangular cards, $ABCD$ and $A'B'C'D'$, which are of different sizes. Put the smaller one onto the bigger one so that two edges of the cards coincide (as shown in Fig.15.6), where $AB = a$, $AD = b$, $AB' = \lambda a$, $AD' = \mu b$. Suppose two points P and Q are on the small card and point R is on the big card.

Prove: $\triangle PQR \leq \frac{1}{2} ab(\lambda + \mu - \lambda\mu)$.

Proof: Using one of A, B, C, D to replace R and two of A, B', C', D' to replace P and Q, we can get some triangles, the biggest one of which is $\triangle D'B'C$. It is easy to figure out that:

$$\triangle D'B'C = \square ABCD - \triangle AB'D' - \triangle B'BC - \triangle DD'C$$

$$= \frac{1}{2}ab(\lambda + \mu - \lambda\mu)$$

The following problem is a variant of Ex15.2.

[Ex15.4] (Moscow Mathematics Competition,1964) Take any 101 points in the interior or on the circumference of a square, where the length of the square's side is 1 and any three points are not collinear.

Prove: you can always find three points such that the area of the triangle defined by the three points is less than $\frac{1}{100}$.

This problem is a little more twisted than Ex15.2. First, we divide the square equally into 50 small rectangles. Then, there exists one rectangle such that at least 3 points of the 101 points are in the interior of or on the circumference of it. So, the area of the triangle defined by the three points is not greater than half of the area of the rectangle, that is $\frac{1}{100}$.

Ex15.2, Ex15.3, Ex15.4 are all concerned with triangles in a quadrilateral. There are some problems which are concerned with the area of quadrilaterals in a triangle. For example:

[Ex15.5] (Poland Mathematics Competition, 1962 ~1963) Cut out a rectangle with the biggest area from a given triangle.

The difficulty of this problem is that there is no limitation on the position of the rectangle. Quite a few people feel that there exists so much variation that they don't know where to start. In some exercise books, the answer to this question is more than 3 pages[2]. Here, a simple method is used to give the general resolution.

---

[2] For example, *Poland Mathematics Competition Exercise Book*, Author: Ye, Boluoyefu, *et al.* Knowledge Publication, 1982

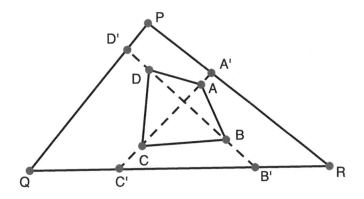

FIGURE 15.7

Proposition 15.2 If there is a convex quadrilateral $ABCD$ in $\triangle PQR$, then at least one of $\triangle BCD$, $\triangle ACD$, $\triangle ABD$ and $\triangle ABC$ is less than or equal to a quarter of $\triangle PQR$ (The problem includes a question of the first time National Middle Student Math Winter Camp Competition).

Proof: Suppose A, B, C, D are on the perimeter of $\triangle PQR$. It's valid since, as shown in Fig.15.7, we can otherwise replace A, B, C, D with A', B', C', D', which are the intersection points of lines AC, BD and the perimeter of $\triangle PQR$. Obviously, we have

$$\triangle ABC \leq \triangle A'B'C'$$

$$\triangle ABD \leq \triangle A'B'D'$$

and so on.

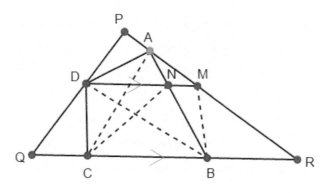

FIGURE 15.8

As shown in the Fig.15.8, point *D* is closer to line *QR* than point *A*. Draw a line parallel to *QR* through *D* which intersects *AB* at *N* and *PR* at *M*. Compare $\triangle NCD$, $\triangle ACD$, and $\triangle BCD$. Since *ABCD* is a convex quadrilateral and *AB* does not intersect *CD*, then at least one of $\triangle ACD$, $\triangle BCD$ is not bigger than $\triangle NCD$.Thus, it is sufficient to prove

$$\triangle NCD \leq \frac{1}{4} \triangle PQR$$

But since $\triangle NCD = \triangle NBD \leq \triangle DMB$, we only need to prove the following proposition:

Proposition 15.3 Suppose there are three points *D, B M* on sides *PQ, QR, RP* of $\triangle PQR$ respectively. If DM ∥ QR , then we have

$$\triangle DMB \leq \frac{1}{4} \triangle PQR$$

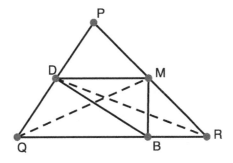

FIGURE 15.9

Proof: As shown in Fig.15.9, we have

$$\frac{\triangle PQR}{\triangle DMB} = \frac{\triangle PQM + \triangle RQM}{\triangle DMB} = \frac{\triangle PQM}{\triangle QDM} + \frac{\triangle RQM}{\triangle RDM}$$

$$= \frac{PQ}{DQ} + \frac{PQ}{PD} = 1 + \frac{PD}{DQ} + 1 + \frac{DQ}{PD} \geq 4$$

Here, the algebraic inequality is used, when a and b are both positive numbers:

$$\frac{b}{a} + \frac{a}{b} \geq 2$$

This is because $(a - b)^2 \geq 0$ and $a^2 + b^2 \geq 2ab$, thus,

$$\frac{b}{a} + \frac{a}{b} = \frac{a^2 + b^2}{ab} \geq 2$$

Now it is not difficult to solve Ex.15.5. With **Proposition 15.2,** the area of the rectangle inscribed in a triangle is not more than half of the area of the triangle. We just need to cut out a rectangle whose area is half of the area of the triangle. Suppose $\angle Q$ and $\angle R$ are not obtuse angles in $\triangle PQR$, and then take the midpoints M and N on PQ and PR respectively. Then cut out the rectangle $GHNM$ as shown in Fig.15.10.

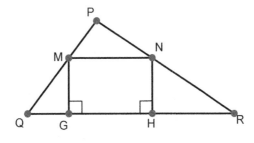

FIGURE 15.10

Propositions 15.1, 15.2 and 15.3 are very useful. We use them frequently when comparing the area of triangles and quadrilaterals.

[Ex15.6] (Mathematics Competitions of Mathematics Amateurs, China, 1982) Take any four points P_1, P_2, P_3, P_4 in the interior or on the circumference of a triangle. If the areas of $\triangle P_i P_j P_k$ (there are four such triangles) are all bigger than $\frac{1}{4} \triangle ABC$,

Prove: one of these four triangles must have an area greater than

$$\frac{3}{4} \triangle ABC$$

With **Proposition 15.2**, Ex15.6 will be smoothly solved. From **Proposition 15.2**, we know that the points P_1, P_2, P_3, P_4 cannot constitute a convex quadrilateral. Hence, there must be a point which is in the interior of a triangle constituted by other 3 points. Let P_4 be in the interior of triangle $P_1 P_2 P_3$, then we have

$$\triangle P_1 P_2 P_3 = \triangle P_1 P_2 P_4 + \triangle P_2 P_3 P_4 + \triangle P_3 P_1 P_4 \quad > \frac{3}{4} \triangle ABC$$

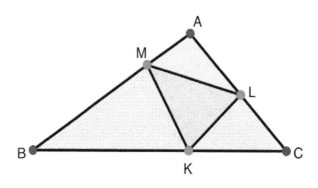

FIGURE 15.11

[Ex15.7] (The 8th International Olympic Mathematics Competition, 1966) In $\triangle ABC$, take three points M, K, L on the sides AB, BC, CA respectively (as shown in Fig.15.11).

Prove: One of the triangles, $\triangle LAM$, $\triangle MBK$ or $\triangle KCL$ must have an area not greater than $\frac{1}{4} \triangle ABC$.

Proof: Let $AM = \lambda AB, BK = \mu BC, CL = \rho CA$. Then,

$$BM = (1 - \lambda)AB$$

$$CK = (1 - \mu)BC$$

$$AL = (1 - \rho)AC$$

By the Co-Angle Theorem, we have

$$\frac{\triangle LAM}{\triangle ABC} \cdot \frac{\triangle MBK}{\triangle ABC} \cdot \frac{\triangle KCL}{\triangle ABC} = \frac{AL \cdot AM}{AC \cdot AB} \cdot \frac{BM \cdot BK}{AB \cdot BC} \cdot \frac{CK \cdot CL}{BC \cdot CA}$$

$$= (1 - \rho) \cdot \lambda \cdot (1 - \lambda) \cdot \mu \cdot (1 - \mu) \cdot \rho$$

$$= \lambda(1 - \lambda) \cdot \mu(1 - \mu) \cdot \rho(1 - \rho)$$

$$\leq \frac{1}{4} \cdot \frac{1}{4} \cdot \frac{1}{4} = \left(\frac{1}{4}\right)^3$$

So, one of these three ratios must not be greater than a quarter.

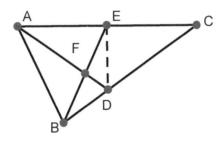

FIGURE 15.12

[Ex15.8] (United States Olympic Mathematics Preliminary Test, 1980) In $\triangle ABC$, $\angle CBA = 72°$ and E is the midpoint of AC. D is on BC and $2BD = DC$. AD intersects BE at F. Then, the ratio of the area of $\triangle BDF$ and that of the quadrilateral $FDCE$ is:

(A) $\frac{1}{5}$, (B) $\frac{1}{4}$, (C) $\frac{1}{3}$, (D) $\frac{2}{5}$, (E) None is correct

Solution: As shown in Fig.15.12, by Co-Angle Theorem and Co-Side Theorem, we have

$$\frac{\triangle BCE}{\triangle BDF} = \frac{BC \cdot BE}{BD \cdot BF}$$

155

$$= 3\left(\frac{BF + FE}{BF}\right) = 3\left(1 + \frac{FE}{BF}\right) = 3(1 + \frac{\triangle ADE}{\triangle ABD})$$

$$= 3\left(1 + \frac{\triangle ADE}{\triangle ADC} \cdot \frac{\triangle ADC}{\triangle ABD}\right) = 3\left(1 + \frac{1}{2} \cdot \frac{2}{1}\right) = 6$$

That is $\triangle BCE$ is 6 times $\triangle BDF$. So $\triangle BDF$ is $\frac{1}{5}$ of quadrilateral $FDCE$. Thus, we choose (A).

It is interesting that the condition $\angle CBA = 72°$ is not used. If you take it into consideration, it will point you in the wrong direction.

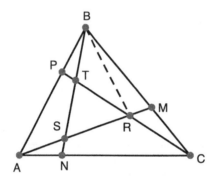

FIGURE 15.13

[Ex15.9] (Poland Mathematics Competition, 1951-1952) As shown in Fig.15.13, take point M, N, P on sides BC, CA, AB of $\triangle ABC$ respectively such that

$$\frac{BM}{MC} = \frac{CN}{NA} = \frac{AP}{PB} = k$$

where $k > 1$ is given. Suppose $\triangle ABC = s$, and then determine the area of $\triangle RST$ formed by the lines AM, BN, CP.

Solution: Using the method of Ex2.3, we have

$$\frac{\triangle ABC}{\triangle ARC} = \frac{\triangle ARC + \triangle CRB + \triangle ARB}{\triangle ARC}$$

$$= 1 + \frac{PB}{AP} + \frac{BM}{MC}$$

$$= 1 + \frac{1}{k} + k$$

$$= \frac{1 + k + k^2}{k}$$

that is,

$$\triangle ARC = \frac{ks}{1 + k + k^2}$$

Likewise both $\triangle BTC$ and $\triangle ASB$ are $\dfrac{ks}{1 + k + k^2}$. So we have

$$\triangle RST = \triangle ABC - \triangle ARC - \triangle BTC - \triangle ASB$$

$$= s - \frac{3ks}{1 + k + k^2} = \frac{(1 - k)^2 s}{1 + k + k^2}$$

In different mathematics competitions, the style of problems is different. Some depend on techniques while some depend on basic skills and straight calculation. For example:

[Ex15.10] (International Olympics Mathematics Competition, 1961) Suppose three sides and the area of a triangle are a, b, c and s.

Prove: $a^2 + b^2 + c^2 \geq 4\sqrt{3}s$.

Proof: By Heron's Formula, we know

$$16s^2 = 4b^2c^2 - \left(b^2 + c^2 - a^2\right)^2$$

So it is sufficient to prove

$$\left(a^2 + b^2 + c^2\right)^2 \geq 3\left[4b^2c^2 - \left(b^2 + c^2 - a^2\right)^2\right]$$

Consider the difference between the left-hand side and the right-hand side:

$$\left(a^2 + b^2 + c^2\right)^2 - 3\left[4b^2c^2 - \left(b^2 + c^2 - a^2\right)^2\right]$$

$$= a^4 + b^4 + c^4 + 2a^2b^2 + 2b^2c^2 + 2a^2c^2 - 12b^2c^2$$

$$+3(b^4 + c^4 + a^4 - 2a^2 b^2 - 2a^2 c^2 + 2b^2 c^2)$$

$$= 4(a^4 + b^4 + c^4 - a^2 b^2 - b^2 c^2 - c^2 a^2)$$

$$= 2[(a^2 - b^2)^2 + (b^2 - c^2)^2 + (c^2 - a^2)^2] \geq 0$$

The proposition is proved.

[Ex15.11] (The 26th Moscow Mathematics Competition, 1963) Suppose A', B', C', D', E' are the midpoints of sides of convex pentagon, respectively.

Prove: The area of pentagon $A'B'C'D'E'$ is more than a half of $ABCDE$.

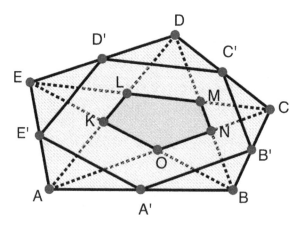

FIGURE 15.14

Proof: As shown in Fig.15.14, connect the diagonals *AC, CE, EB, BD, DA* of pentagon *ABCDE*, then five triangles$\triangle ABC, \triangle BCD, \triangle CDE, \triangle DEA \triangle EAB$ cannot cover the Pentagon twice, so

$$\triangle ABC + \triangle BCD + \triangle CDE + \triangle DEA + \triangle EAB < 2S_{ABCDE}$$

$$\therefore \triangle BB'A' + \triangle CC'B' + \triangle DD'C' + \triangle EE'D' + \triangle AA'E'$$

$$= \frac{1}{4}\triangle ABC + \frac{1}{4}\triangle BCD + \frac{1}{4}\triangle CDE$$

$$+ \frac{1}{4}\triangle DEA + \frac{1}{4}\triangle EAB < \frac{1}{2}S_{ABCDE}$$

$$\therefore S_{A'B'C'D'E'} > \frac{1}{2} S_{ABCDE}$$

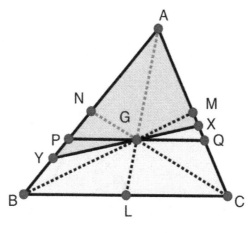

FIGURE 15.15

[Ex15.12] (Mathematics Competition in Anhui Province, China, 1978) Draw a line through the centroid of a triangle, which divides the triangle into two parts.

Prove: The difference between the area of the two parts is not more than $\frac{1}{9}$ of the area of the triangle.

Proof: As shown in Fig.15.15, suppose the centroid of $\triangle ABC$ is G, and that three medians are AL, BM, CN. Draw a line parallel to BC through G intersecting AB and AC at P and Q, respectively. A line through G intersects AB and AC at Y, X, respectively where Y lies between B and P, and X lies between M and Q. It is obvious that $\triangle APQ = \frac{4}{9} \triangle ABC$, $\triangle BCM = \frac{1}{2} \triangle ABC$, so it is sufficient to prove

$$\triangle APQ \leq \triangle AXY \leq \triangle ABM$$

Since

$$\frac{\triangle XGQ}{\triangle YGP} = \frac{XG \cdot QG}{YG \cdot PG} = \frac{XG}{YG} = \frac{\triangle AXG}{\triangle AYG} \leq 1$$

that is

$$\triangle YGP \geq \triangle XGQ, \qquad \triangle AXY \geq \triangle APQ = \frac{4}{9} \triangle ABC$$

Also
$$\frac{\triangle XGM}{\triangle YGB} = \frac{GX \cdot GM}{GY \cdot GB} = \frac{1}{2}\frac{GX}{GY} \leq \frac{1}{2}$$

That is
$$\triangle YGB \geq \triangle XGM$$

So
$$\triangle AXY \leq \triangle ABM = \frac{1}{2}\triangle ABC .$$

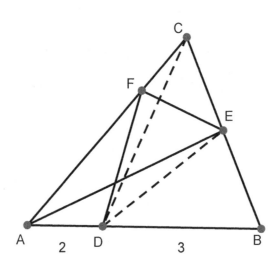

Figure 15.16

[Ex15.13] (United States High School Mathematics Competition, 1983) As shown in Fig.15.16, the area of $\triangle ABC$ is 10, D, E, F are on sides AB, BC, CA respectively, and AD = 2, DB = 3. If the area of $\triangle ABE$ is equal to the area of quadrilateral $DBEF$, then the area is:

$(A)4,$ $(B)5,$ $(C)6,$ $(D)\frac{5}{3}\sqrt{10},$ $(E)not\ determined\ uniquely$

Solution:
$$\because \triangle ABE = S_{DBEF}$$

$$\therefore \triangle ADE + \triangle BDE = \triangle FDE + \triangle BDE$$

$$\therefore \triangle ADE = \triangle FDE$$

160

$$\therefore AF \parallel DE$$

$$\therefore \triangle ABE = \triangle ADE + \triangle BDE$$

$$= \triangle CDE + \triangle BDE = \triangle CBD$$

$$= \frac{\triangle CBD}{\triangle ABC} \cdot \triangle ABC = \frac{BD}{AB} \cdot \triangle ABC$$

$$= \frac{3}{5} \times 10 = 6$$

So, we choose C.

The relationship between the area and the parallel is very useful in solving problems. The following is another example:

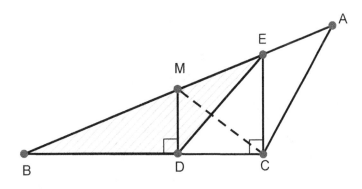

FIGURE 15.17

[Ex15.14] (United States High School Mathematics Competition, 1984) As shown in Fig.15.17, in an obtuse triangle $\triangle ABC$, $AM = MB$, $MD \perp BC$, $EC \perp BC$. If $\triangle ABC = 24$, then the area of $\triangle BED$ is:

(A)9 (B)12 (C)15 (D)18 (E)not determined uniquely

Solution:

$$\because MD \perp BC, EC \perp BC$$

$$\therefore MD \parallel EC$$

$$\therefore \triangle MDE = \triangle MDC$$

$$\therefore \triangle BED = \triangle BMD + \triangle MDE$$

$$= \triangle BMD + \triangle MDC = \triangle BMC = \frac{1}{2} \triangle ABC$$

$$= 12$$

So, we choose B.

In some books (such as *China and United States Mathematics Competitions Solutions*, Shanghai Science and Technology Press, 1987), the solutions are very complicated because one does not use the relationship of the parallel and area, but uses similar shape and trigonometric functions instead.

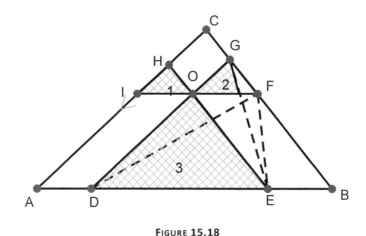

FIGURE 15.18

[Ex15.15] (The 2nd United States Mathematics Invitational Competition, 1984) As shown in Fig.15.18, take any point O in the interior of $\triangle ABC$, and draw three lines through O which are parallel to sides AB, BC, AC. The areas of triangles $\triangle 1$, $\triangle 2$, $\triangle 3$ are 4, 9, 49 respectively. **Determine** the area of $\triangle ABC$.

Solution: Let $GO = \lambda GD$, and then $DO = (1 - \lambda)GD$.

So
$$\frac{9}{49} = \frac{\triangle_2}{\triangle_3} = \frac{\triangle_2}{\triangle DOF} \cdot \frac{\triangle DOF}{\triangle EOF} \cdot \frac{\triangle GOE}{\triangle_3}$$

$$= \frac{GO}{DO} \cdot \frac{1}{1} \cdot \frac{GO}{DO} = \frac{\lambda^2}{(1-\lambda)^2}$$

$$\therefore \triangle EOF = \triangle GOE \cdot \frac{\triangle_3}{\triangle_3} = \frac{GO}{DO} \triangle_3$$

$$= \frac{\lambda}{1-\lambda} \triangle_3 = \sqrt{\frac{9}{49}} \triangle_3 = 21$$

Likewise $\triangle IDO = 14, \quad \triangle HOG = 6.$

So

$$\triangle ABC = \triangle_1 + \triangle_2 + \triangle_3$$

$$+ 2(\triangle EOF + \triangle IDO + \triangle HOG) = 144$$

ADDITIONAL PROBLEMS

[P15.1] Suppose $\triangle PAB$ and $\triangle QAB$ are inscribed triangles of $\odot O$.

Given: $\triangle PAB > \triangle QAB$

Think about whether there must exist the following inequality:

$$\frac{\triangle PAB}{\triangle QAB} > \frac{PA + PB + AB}{QA + QB + AB}$$

[P15.2] Suppose $\triangle PQR$ is an inscribed triangle in convex quadrilateral $ABCD$.

Prove: For $\triangle ABC$, $\triangle ABD$, $\triangle ACD$ and $\triangle BCD$, at least of these triangles must have an area greater than or equal to the area of $\triangle PQR$.

163

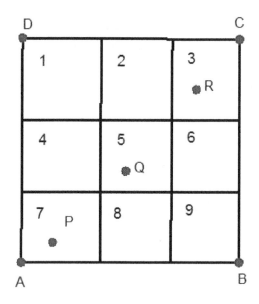

FIGURE 15.19

[P15.3] As shown in Fig.15.19, divide square *ABCD* into 9 small squares of equal area. Take points *P, Q, R* from the interior or circumference of three small squares labeled 7, 5 and 3 respectively. What is the largest area of $\triangle PQR$?

[P15.4] A point *P* is on an equilateral triangular card, and the distance from *P* to three sides of the equilateral triangle are 3cm, 5cm, 7cm. What is the minimum area of a triangle cut out through *P*?

[P15.5] Take four points *P, Q, R, S*, one on each of four sides *AB, BC, CD, DA* of parallelogram *ABCD* such that $AP = 2PB, BQ = 2QC, CR = 2RD, DS = 2SA$. What is the ratio of the area of the convex quadrilateral enclosed by *PC, QD, RA, SB* to the area of *ABCD*?

[P15.6] (United States High School Mathematics Competition, 1985) The diagonal *BD* of the rectangle *ABCD* is divided equally into three segments by two points, *P* and *Q*. The length of each of the three segments of *BD* are all equal to 1. The two lines *AP* and *CQ* are perpendicular to *BD*. Then the area of *ABCD*, rounded to one decimal place, is:

$(A)4.1$ $\qquad$ $(B)4.2$ $\qquad$ $(C)4.3$ $\qquad$ $(D)4.4$ $\qquad$ $(E)4.5$

[P15.7] (United States Mathematics Invitational Competition, 1985) Suppose *ABCD* is a unit square. Take four points *P, Q, R, S* on sides *AB, BC, CD, DA* respectively such that

$$AP = BQ = CR = DS = \frac{1}{n}AB$$

If the area of the square delimited by lines *AR, BS, CP, DQ* is $\frac{1}{1985}$, determine *n*.

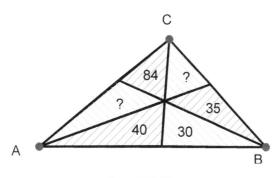

FIGURE 15.20

[P15.8] (United States Mathematics Invitational Competition, 1985) As shown in Fig.15.20, three lines intersecting at one point divide $\triangle ABC$ into 6 small triangles. Given the area of four triangles which have been labeled in figure 15.20, determine the area of $\triangle ABC$.

[P15.9] (High School Mathematics League in Provinces, Municipalities and Autonomous Regions, 1984) In $\triangle ABC$, point *P* is on side *BC*. Take two points *F* and *E* on sides *AB* and *AC* such that *AFPE* is a parallelogram.

Prove: Of $\triangle BPF, \triangle CPE, \square AFPE$, there must exist at least one whose area is greater than or equal to $\frac{4}{9}$ of $\triangle ABC$.

CHAPTER 16

USING THE AREA METHOD TO SOLVE MATHEMATICS COMPETITION PROBLEMS

Most plane geometric problems in mathematics competitions can be solved by the area method. The following examples are some choices to ponder.

[Ex16.1] (International Olympics Mathematics Competition, 1959)

Given: hypotenuse c, try constructing a right triangle such that the median on the hypotenuse is the geometric mean of the two legs.

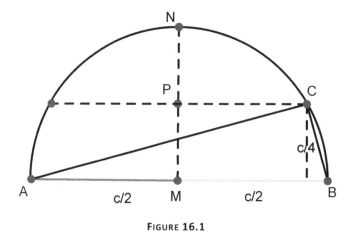

FIGURE 16.1

Analysis: Let $\triangle ABC$ be a right triangle, based on the condition $AB = c$, a half-circle with diameter AB is constructed. Then point C must lie on this half-circle. In addition, the

median on the hypotenuse should be $\dfrac{C}{2}$. So based on the given we have

$AC \cdot BC = \left(\dfrac{C}{2}\right)^2 = \dfrac{c^2}{4}.$ Hence $\triangle ABC = \dfrac{1}{2}AC \cdot BC = \dfrac{1}{8}c^2,$ which means that the

height of the hypotenuse is $\dfrac{c}{4},$ so we have the method of construction.

Construction: Draw a circle whose center is M and radius is $\dfrac{c}{2}.$ And suppose AB is

the diameter of⊙M. Draw a perpendicular to AB through M, intersecting the circle at point N. Take the midpoint of MN labeled P, and draw a parallel line to AB through P intersecting ⊙M at point C. Then the triangle $\triangle ABC$ satisfies all given conditions.

Proof:
$$AC \cdot BC = 2\triangle ABC = \dfrac{2PM \cdot AB}{2} = \dfrac{MN}{2} \cdot c = \left(\dfrac{c}{2}\right)^2$$

[Ex16.2] (International Olympics Mathematics Competition, 1961) A point P is in the interior $\triangle P_1 P_2 P_3$. $P_1 P$, $P_2 P$ and $P_3 P$ intersect their opposite sides at Q_1, Q_2 and Q_3, respectively, as shown in Fig.16.2.

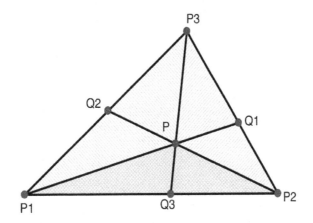

FIGURE 16.2

Prove: Of the 3 ratios $\dfrac{P_1 P}{PQ_1},$ $\dfrac{P_2 P}{PQ_2}$ and $\dfrac{P_3 P}{PQ_3}$ there must be at least one that is

less than or equal to 2, and one that is greater than or equal to 2.

Proof: (Based on the Co-Side Theorem)

$$\frac{PQ_1}{P_1Q_1} + \frac{PQ_2}{P_2Q_2} + \frac{PQ_3}{P_3Q_3} = \frac{\Delta PP_2P_3 + \Delta PP_3P_1 + \Delta PP_1P_2}{\Delta P_1P_2P_3} = 1$$

So among $\dfrac{P_1P}{PQ_1}, \dfrac{P_2P}{PQ_2}$ and $\dfrac{P_3P}{PQ_3}$, there must be at least one that is less than or equal to $\dfrac{1}{3}$, and one that is greater than or equal to $\dfrac{1}{3}$.

[Ex16.3] (International Olympics Mathematics Competition, 1962) There is an isosceles triangle whose circumradius is R and the radius of its inscribed circle is r.

Prove: The distance of centers of the two circles is $\sqrt{R(R-2r)}$

Solution: Suppose that the height of this isosceles triangle is h, the base is $2b$ and two diagonals are length a. Based on the Pythagorean Theorem, we have:

$$a = \sqrt{b^2 + h^2}$$

Based on the calculation of area, we have:

$$bh = \Delta = \frac{1}{2}r(2a + 2b) = r(a + b)$$

$$\because r = \frac{bh}{a + b} = \frac{bh(a - b)}{a^2 - b^2} = \frac{b}{h}\left(\sqrt{b^2 + h^2} - b\right)$$

$$2R = \frac{a^2}{h} = \frac{b^2 + h^2}{h}$$

$$\therefore R = \frac{b^2 + h^2}{2h}$$

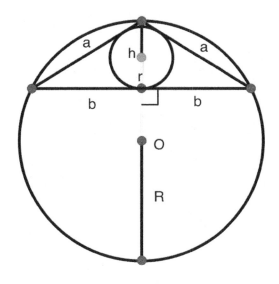

FIGURE **16.3**

As shown in Fig 16.3, the distance between two centers is

$$R - h + r = \frac{b^2 + h^2}{2h} - h + \frac{b\sqrt{b^2 + h^2}}{h} - \frac{b^2}{h}$$

$$= \frac{b^2}{2h} - \frac{h}{2} - \frac{b^2}{h} + \frac{b\sqrt{b^2 + h^2}}{h}$$

$$= \frac{-(b^2 + h^2) + 2b\sqrt{b^2 + h^2}}{2h}$$

On the other hand

$$R(R - 2r) = \frac{b^2 + h^2}{2h}\left[\frac{b^2 + h^2}{2h} - \frac{2b}{h}\left(\sqrt{b^2 + h^2} - b\right)\right]$$

$$= \frac{(b^2 + h^2)^2 - 4b(b^2 + h^2)\sqrt{b^2 + h^2} + 4b^2(b^2 + h^2)}{4h^2}$$

$$= \left[\frac{-(b^2 + h^2) + 2b\sqrt{b^2 + h^2}}{2h}\right]^2 = (R - h + r)^2$$

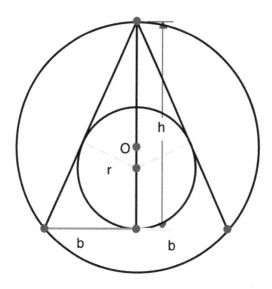

FIGURE 16.4

Here, what we need to explain is, as shown in Fig.16.4, this distance between two centers should be

$$h - R - r = -(R - h + r)$$

The results of calculating are the same.

There is another case:

As shown in Fig.16.5, the distance is:

$$r - (OD) = r - (h - R) = r + R - h$$

The result is the same as in Fig.16.3.

There isn't any special skill to solve this question. And as long as you calculate it step by step, you can obtain success. This, then can test the basic skills and carefulness of contestants.

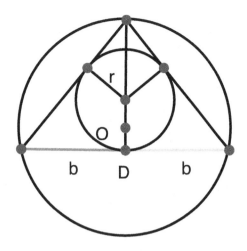

FIGURE 16.5

[Ex16.4] (Rewrite based on the 5th question in International Olympics Mathematics Competition, 1965)[3]

Given: In ΔOAB, $\angle O$ is an acute angle, and AC, BD are altitudes. Take any point M on AB, and draw the perpendicular segments MP and MQ to lines OA and OB respectively. Then draw perpendicular segments PK to OB and QT to OA (fig. 16.6)

Prove: The intersection point of PK and QT lies on line segment CD.

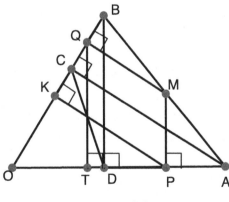

FIGURE 16.6

Solution: To prove the intersection point of PK and QT lies on CD, is just to prove PK, QT and CD are concurrent. So we can suppose QT intersects CD at point S, then prove

---

[3] in the 5th question we are asked: when M moves on AB, what's the locus of the intersection point of PK and QT?

PK goes through point S. Because PK is perpendicular to OB, it is sufficient for us to prove PS is perpendicular to OB, i.e. $PS \parallel AC$ or $PS \parallel MQ$, which means that we should prove: $\Delta MQS = \Delta MQP$ or $\Delta ACP = \Delta ACS$

In the following, we will prove $\Delta ACP = \Delta ACS$:

$$\Delta ACS = \frac{\Delta ACS}{\Delta ACD} \cdot \frac{\Delta ACD}{\Delta ACP} \cdot \Delta ACP = \frac{CS}{CD} \cdot \frac{AD}{AP} \cdot \Delta ACP$$

$$= \frac{\Delta QCT \cdot \Delta ACP}{\Delta QCD + \Delta QTD} \cdot \frac{\Delta AMD}{\Delta AMP}$$

$$= \frac{\Delta QCT \cdot \Delta ACP}{\Delta QCD + \Delta QTB} \cdot \frac{\Delta APB}{\Delta AMP}$$

$$= \frac{\Delta QCT}{\Delta BCT} \cdot \frac{AB}{AM} \cdot \Delta ACP$$

$$= \frac{QC}{BC} \cdot \frac{AB}{AM} \cdot \Delta ACP$$

$$= \frac{\Delta AQC}{\Delta ABC} \cdot \frac{\Delta ABC}{\Delta AMC} \cdot \Delta ACP$$

$$= \frac{\Delta AQC}{\Delta AMC} \cdot \Delta ACP = \Delta ACP$$

[Ex16.5] (International Olympics Mathematics Competition, 1970) **Given:** M is an any point on side AB of ΔABC. r, r_1 and r_2 are the radii of inscribed circles ΔABC, ΔBCM and ΔACM respectively. And ρ, ρ_1 and ρ_2 are the radii of the escribed circles of ΔABC, ΔBCM and ΔACM respectively. All of these escribed circles are in the interior of $\angle ACB$.

Prove:

$$\frac{r}{\rho} = \frac{r_1}{\rho_1} \cdot \frac{r_2}{\rho_2}$$

Proof: Label the three sides of ΔABC as a, b and c. And suppose $CM = l$, AM=c_1 and $MB = c_2$. Based on the area calculation, we have

$$\frac{1}{2}r(a+b+c) = \Delta ABC = \frac{1}{2}\rho(a+b-c)$$

$$\therefore \frac{r}{\rho} = \frac{a+b-c}{a+b+c} \tag{16.1}$$

Similarly, we have

$$\frac{r_1}{\rho_1} = \frac{b+l-c_1}{b+l+c_1}, \qquad \frac{r_2}{\rho_2} = \frac{a+l-c_2}{a+l+c_2} \tag{16.2}$$

$$\therefore \frac{r_1}{\rho_1} \cdot \frac{r_2}{\rho_2} = \frac{l^2+l(b+a-c_1-c_2)+(b-c_1)(a-c_2)}{l^2+l(b+a+c,+c_2)+(b+c_1)(a+c_2)}$$

By using $c_1 + c_2 = c$ and the Pythagorean Difference Theorem, we have:

$$\frac{c_1}{c} = \frac{\Delta AMC}{\Delta ABC} = \frac{b^2+c_1^2-l^2}{b^2+c^2-a^2}$$

$$l^2 = b^2 + c_1^2 - \frac{c_1(b^2+c^2-a^2)}{c}$$

is solved, and by plugging it into the equation above and then rearranging it, we have:

$$\frac{r_1}{\rho_1} \cdot \frac{r_2}{\rho_2} = \frac{[bc+cl+c_1(a+b)](a+b-c)}{[bc+cl+c_1(a-b)](a+b+c)}$$

$$= \frac{a+b-c}{a+b+c} = \frac{r}{\rho}$$

This example tells us that when proving some questions whose formulas seem very complicated, we shouldn't be afraid of them. As long as we calculate them carefully, most of them can be eliminated.

In fact, there is a smart way to solve this question, as shown in Fig.16.7.

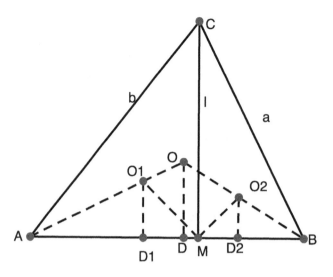

FIGURE 16.7

Suppose O is the intersection point of the bisectors of $\angle A$ and $\angle B$. O_1 is the intersection point of AO and the bisector of $\angle AMC$. O_2 is the intersection point of BO and the bisector of $\angle BMC$. Then point O, $O1$ and $O2$ are the incenters of $\triangle ABC$, $\triangle AMC$ and $\triangle BMC$ respectively. Draw three perpendiculars to AB: OD, $O1D1$ and $O2D2$. Let $r = OD$, $r_1 = O_1D_1$, $r_2 = O_2D_2$. Suppose $s = \frac{1}{2}(a + b + c)$, Based on (16.1) and Heron's Formula, we have:

$$\frac{r}{\rho} = \frac{a + b - c}{a + b + c} = \frac{s - c}{s}$$

$$= \frac{1}{s} \cdot \frac{\Delta^2}{s(s - a)(s - b)} = \frac{\Delta^2}{s^2(s - a)(s - b)}$$

$$= \frac{r^2}{(s - a)(s - b)} = \frac{OD}{AD} \cdot \frac{OD}{BD} = \frac{O_1D_1}{AD_1} \cdot \frac{O_2D_2}{BD_2}$$

Here, we used $\Delta^2 = s(s - a)(s - b)(s - c)$, $AD = s - a$, and $BD = s - b$.

Similarly:

$$\frac{r_1}{\rho_1} = \frac{r_1^2}{AD_1 \cdot D_1 M} = \frac{O_1 D_1}{AD_1} \cdot \frac{O_1 D_1}{D_1 M}$$

$$\frac{r_2}{\rho_2} = \frac{r_2^2}{MD_2 \cdot D_2 B} = \frac{O_2 D_2}{MD_2} \cdot \frac{O_2 D_2}{D_2 B}$$

Notice that $\angle O_1 M O_2 = 90°$, So $\Delta O_1 D_1 M \sim \Delta M D_2 O_2$, thus

$$\frac{O_1 D_1}{D_1 M} \cdot \frac{O_2 D_2}{D_2 M} = 1$$

Thus we have:

$$\frac{r_1}{\rho_1} \cdot \frac{r_2}{\rho_2} = \frac{O_1 D_1}{AD_1} \cdot \frac{O_1 D_1}{D_1 M} \cdot \frac{O_2 D_2}{D_2 M} \cdot \frac{O_2 D_2}{D_2 B}$$

$$\frac{O_1 D_1}{AD_1} \cdot \frac{O_2 D_2}{D_2 B} = \frac{OD}{AD} \cdot \frac{OD}{BD} = \frac{r}{\rho}$$

[Ex16.6] (International Olympics Mathematics Competition, 1990) In a circle, chord AB intersects chord CD at point E. Take a point M on line segment BE and draw the circumcircle of ΔDEM and its tangent line through E such that the tangent intersects chord BC at F and the extension line of CA at G. As shown in Fig.16.8.

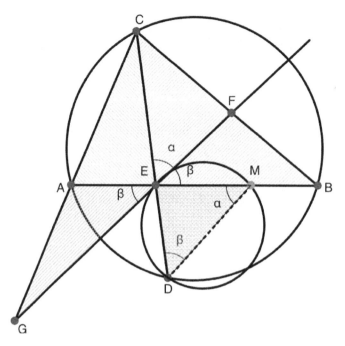

FIGURE 16.8

Given:

Determine the ratio

Solution: Notice that

$$\angle BEF = \angle AEG = \angle MDE = \beta$$

$$\angle CEF = \angle DME = 180° - \angle CEG = \alpha$$

Apply the Co-Side Theorem and Co-Angle Theorem, we have:

$$\frac{AE}{BE} = \frac{\triangle AEC}{\triangle BEC} = \frac{\triangle GEC - \triangle GEA}{\triangle FEC + \triangle FEB}$$

$$= \frac{\dfrac{\triangle GEC - \triangle GEA}{\triangle MDE}}{\dfrac{\triangle FEC + \triangle FEB}{\triangle MDE}}$$

$$= \frac{\dfrac{GE \cdot CE}{ME \cdot MD} - \dfrac{GE \cdot AE}{DE \cdot MD}}{\dfrac{FE \cdot CE}{ME \cdot MD} + \dfrac{FE \cdot BE}{DE \cdot MD}}$$

$$= \frac{GE}{FE} \cdot \left(\frac{CE \cdot DE - ME \cdot AE}{CE \cdot DE + ME \cdot BE} \right)$$

$$= \frac{GE}{FE} \cdot \left(\frac{AE \cdot BE - ME \cdot AE}{AE \cdot BE + ME \cdot BE} \right)$$

(By the Intersecting Chord Theorem : $CE \cdot DE = AE \cdot BE$)

$$= \frac{AE}{BE} \cdot \frac{GE}{FE} \cdot \frac{(BE - ME)}{(AE + ME)}$$

$$= \frac{AE}{BE} \cdot \frac{GE}{FE} \cdot \frac{BM}{AM}$$

$$\therefore \frac{GE}{FE} \cdot \frac{BM}{AM} = \lambda$$

$$\therefore \frac{GE}{FE} = \frac{\lambda}{1 + \lambda}$$

[Ex16.7] (International Olympics Mathematics Competition, 1991) Take any point P in the interior of $\triangle ABC$.

Prove: There must be at least one among $\angle PAB$, $\angle PAC$ and $\angle PCA$, which is less than or equal to $30°$.

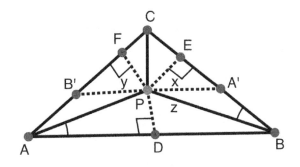

FIGURE 16.9

177

Solution: As shown in the Fig.16..9, it is sufficient to prove that at least one of these three ratios:

$$\frac{PD}{AP}, \frac{PE}{BP}, \frac{PF}{CP}$$

is less than or equal to $\frac{1}{2}$.

Here, PD, PE and PF are perpendicular line segments to sides AB, BC and CA respectively. x, y and z denote the length of PE, PF and PD, respectively. Draw a line through P which intersects BC and AC at point A' and B', respectively such that $\angle CA'B' = \angle CAB$, then $\angle CB'A' = \angle CBA$. So $\triangle ABC \sim \triangle A'B'C$. Let a, b, c and a', b', c', denote the three sides of $\triangle ABC$ and $\triangle A'B'C$ respectively, then there exists a $k > 0$ such that $a' = ka$, $b' = kb$ and $c' = kc$.

Based on the following relationship of areas

$$\triangle PA'C + \triangle PB'C = \triangle A'B'C$$

We have

$$x \cdot A'C + y \cdot B'C = \triangle A'B'C \leq A'B' \cdot PC$$

That is

$$b'x + a'y \leq c'PC$$

$$\therefore kbx + kay \leq kcPC$$

i.e.

$$bx + ay \leq c \cdot PC \Rightarrow \frac{b}{c}x + \frac{a}{c}y \leq PC$$

Similarly

$$cx + az \leq b \cdot PB \Rightarrow \frac{c}{b}x + \frac{a}{b}z \leq PB$$

$$cy + bz \leq a \cdot PA \Rightarrow \frac{c}{a}y + \frac{b}{a}z \leq PA$$

Adding the three inequalities together, we have

$$\left(\frac{b}{c} + \frac{c}{b}\right)x + \left(\frac{a}{c} + \frac{c}{a}\right)y + \left(\frac{a}{b} + \frac{b}{a}\right)z \leq PA + PB + PC$$

$$\therefore 2(x + y + z) \leq PA + PB + PC$$

which means that there must be at least one of $2x \leq PB$, $2y \leq PC$ and $2z \leq PA$ that holds, i.e. at least one of $\dfrac{PD}{AP}$, $\dfrac{PE}{BP}$ and $\dfrac{PF}{CP}$ is less than or equal to $\dfrac{1}{2}$.

It should be noted that we have just proved a very famous geometry inequality:

The Erdos Mundell Inequality: Take a point P from the interior or circumference of $\varDelta ABC$, and label the distance from P to the three sides of $\varDelta ABC$ as x, y and z, respectively, then

$$x + y + z \le \frac{1}{2}(PA + PB + PC)$$

The equal sign in this inequality holds if and only if $\varDelta ABC$ is an equilateral triangle and P is center of $\varDelta ABC$.

In mathematics competitions, some questions are related to some famous theorems or formulas, and the example above is just one of them. The following two examples are also related to famous questions.

[Ex16.8] (Bonus Question of Shanxi High Middle School Mathematics Competition, the second test, China, 1978) A point M is in the interior of an acute triangle ABC such that

$$\angle AMB = \angle BMC = \angle CMA = 120°$$

As shown in Fig.16.10, draw the perpendiculars to MA, MB and MC through A, B and C, respectively. The intersection points these perpendiculars constitute $\triangle DEF$

Any point P is in the interior of DEF.

Prove: $PA + PB + PC \ge MA + MB + MC$ (Fig.16.10)

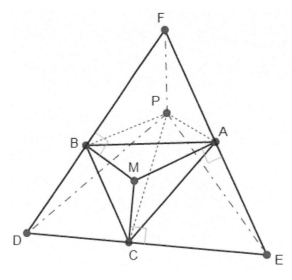

Solution: As shown in Fig.16.10, draw the perpendiculars to MA, MB and MC through A, B and C, respectively. The intersection points these perpendiculars constitute $\triangle DEF$, then we have

$$\angle BDC = 180° - \angle BMC = 180° - 120° = 60°.$$

Similarly, $\angle CEA = 60°$. So $\triangle DEF$ is an equilateral triangle and let $DE = EF = FD = d$. Then

$$PA \cdot d \geq 2 \triangle PEF$$

$$PB \cdot d \geq 2 \triangle PDF$$

$$PC \cdot d \geq 2 \triangle PDE$$

Add these:

$$\therefore \quad (PA + PB + PC)d \geq 2(\triangle PEF + \triangle PDF + \triangle PDE)$$
$$= 2 \triangle DEF$$
$$= 2(\triangle MDE + \triangle MEF + \triangle MFD)$$
$$= (MA + MB + MC)d$$

$$\therefore \quad PA + PB + PC \geq MA + MB + MC$$

This question above is a special case of the famous Fermat Problem, yet the general case is : let A, B and C be three mines, whose productions of ore are x, y and z every day respectively. Now we want to find a place where this ore can be collected. What place can we choose such that the freight cost is smallest? It means that we want to find a point P such that

$$xPA + yPB + zPC$$

is smallest.

It is not difficult to solve this more general problem by a similar method. In the following, we will describe simply the idea and steps to a solution.

First, if the biggest one of x, y and z is greater than or equal to the sum of the other two, for example, $x \geq y + z$, then we choose point A. (Why?)

In the following, suppose the sum of any two of x, y and z is greater than the third one. So we may draw a triangle DEF such that the ratio of its three sides is $x : y : z$, and its three angles are $\angle D$, $\angle E$ and $\angle F$. Then take a point M from the interior of $\triangle ABC$ such that

$$\angle BMC = 180° - \angle D$$

$$\angle CMA = 180° - \angle E$$

$$\angle AMB = 180° - \angle F$$

Then the point M is what we want.

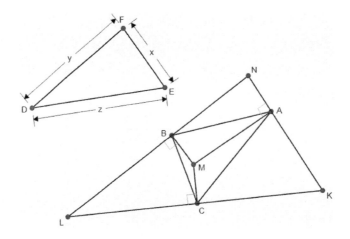

FIGURE 16.11

The proof is similar to Ex16.8: draw perpendiculars to MA, MB and MC through A, B and C, respectively. Let L, K, N be the intersections of these perpendiculars, then $\triangle LKN \sim \triangle DEF$ (As shown in Fig.16.11). For any point P, we have:

$$PA \cdot KN + PB \cdot LN + PC \cdot LK \geq 2 \triangle LKN$$

$$= MA \cdot KN + MB \cdot LN + MC \cdot LK$$

Because $KN:LN:LK = x:y:z$, we have:

$$xPA + yPB + zPC \geq xMA + yMB + zMC.$$

The last problem is how to find point M?

Notice that $\angle AMB = 180^\circ - \angle F$. We can construct it this way: draw a triangle $\triangle ABN'$ at the outside of $\triangle ABC$ such that $\angle AN'B = \angle F$. Now draw the circumcircle

of $\triangle AN'B'$. Then any point M' lying on $\overset{\frown}{AB}$ satisfies $\angle AM'B = 180^\circ - \angle F$. Similarly, draw $\triangle ACK$ such that $\angle AK'C = \angle E$. And then draw circumcircle of $\triangle ACK'$.

The intersection point M (different from point A) satisfies:

$$\angle AMB = 180^\circ - \angle F, \quad \angle AMC = 180^\circ - \angle E.$$

So $\angle BMC = 180^\circ - \angle D$.

182

If the point M is outside of $\triangle ABC$, we can prove that: the objective is minimized at the vertex of the biggest angle. Here we won't go into detail.

[Ex16.9] (High School Student Mathematics League in Provinces, Municipalities and Autonomous Regions, China, 1986) R is the radius of the circumcircle of an acute triangle ABC. Points D, E and F are on sides BC, CA and AB, respectively.

Prove: AD, BE and CF are three altitudes of $\triangle ABC$ if and only if

$$s = \frac{R}{2}(EF + FD + DE)$$

where, s is the area of $\triangle ABC$.

Solution: The proof of necessity: if AD, BE and CF are three altitudes of $\triangle ABC$, Then $\angle BEC = \angle BFC = 90°$. So points B, F, E and C lie on a circle. So:

$$\angle ABC + \angle CEF = 180°$$

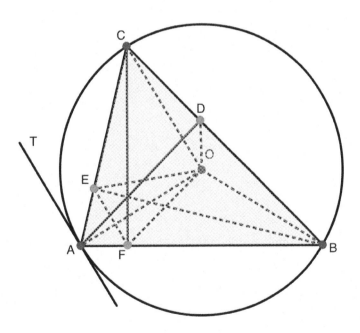

Figure 16.12

Draw a line AT which is tangent with the circumcircle of $\triangle ABC$ at point A (As shown in Fig.16.12). Then:

$$\angle CAT = \angle ABC = 180° - \angle CEF = \angle FEA$$

$$\therefore AT \parallel EF$$

Let the center of circumcircle of $\triangle ABC$ be O, then radius $OA \perp AT$,

i.e. we have $OA \perp EF$

$$\therefore \quad S_{OFAE} = \frac{R}{2} EF$$

Similarly

$$S_{OFBD} = \frac{R}{2} DF, \qquad S_{ODCE} = \frac{R}{2} DE$$

Adding the three equalities above, we have:

$$\triangle ABC = S_{OFAE} + S_{OFBD} + S_{ODCE}$$
$$= \frac{R}{2}(EF + FD + DE)$$

The proof of sufficiency:

Given:
$$\triangle ABC = \frac{R}{2}(EF + FD + DE)$$

we know that $OA \perp EF, \quad OB \perp DF, \quad OC \perp DE$ or else

$$\triangle ABC < \frac{R}{2}(EF + FD + DE)$$

which is contradictory to the supposition.

So $\quad \angle AEF = \angle TAE = \angle ABC.$ Thus B, F, E and C are on a circle.

Similarly, B, D, E and A lie on a circle.

$$\therefore \quad \angle BFC = \angle BEC$$

Similarly $\angle AFC = \angle ADC$

$$\therefore \quad \angle BEC + \angle ADC = \angle BFC + \angle AFC = 180°$$

Based on B, D, E and A lying on a circle, we have $\angle BDA = \angle BEA$,

So $\qquad\qquad \angle BEC = \angle ADC$

$$\therefore \angle BEC = \angle ADC = 90°$$

Similarly, we can prove $\triangle BFC = 90°$.

i.e. $BE \perp AC, CF \perp AB, AD \perp BC$.

From this question, we can find that in all inscribed triangles of an acute triangle, the one with smallest perimeter is the triangle whose vertexes are the feet of perpendiculars of their three altitudes. The reason is the following: If there are other three points x, y and z lying on BC, CA and AB, respectively, then it is impossible that $OA \perp YZ, OB \perp XZ, OC \perp XY$ hold at the same time. So:

$$\frac{R}{2}(EF + FD + DE) = \triangle ABC$$
$$= S_{OFAE} + S_{OFBD} + S_{ODCE}$$
$$< \frac{R}{2}(YZ + ZX + XY)$$
$$\therefore EF + FD + DE < YZ + ZX + XY$$

This is a famous problem in history. But the proof based on the area method may be considered novel.

[Ex16.10] (International Olympics Mathematics Competition, 1980) Let point P lie in the interior of $\triangle ABC$. Point D, E and F are respectively the feet of perpendiculars drawn from P to BC, CA and AB. Calculate all points that make

$$\frac{a}{PD} + \frac{b}{PE} + \frac{c}{PF}$$

smallest. (Fig.16.13)

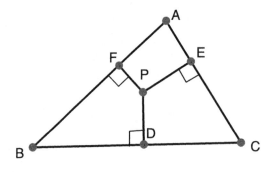

FIGURE 16.13

Solution: Obviously , we have

$$a \cdot PD + b \cdot PE + c \cdot PF = 2 \triangle ABC$$

$$\therefore 2 \triangle ABC \left(\frac{a}{PD} + \frac{b}{PE} + \frac{c}{PF} \right)$$

$$= (a \cdot PD + b \cdot PE + c \cdot PF) \left(\frac{a}{PD} + \frac{b}{PE} + \frac{c}{PF} \right)$$

$$= a^2 + b^2 + c^2 + ab \left(\frac{PD}{PE} + \frac{PE}{PD} \right) + bc \left(\frac{PF}{PE} + \frac{PE}{PF} \right) + ca \left(\frac{PD}{PF} + \frac{PF}{PD} \right)$$

$$\geq a^2 + b^2 + c^2 + 2ab + 2bc + 2ca$$

$$= (a + b + c)^2$$

$$\therefore \frac{a}{PD} + \frac{b}{PE} + \frac{c}{PF} \geq \frac{(a + b + c)^2}{2 \triangle ABC}$$

If and only if $PD = PE = PF$, the equal sign can be chosen in the inequalities above. So P is the incenter of $\triangle ABC$.

In the process of the solution, $\frac{1}{x} + x \geq 2(x > 0)$ is used again. This inequality is very important and used many times above. It is usually used in solving geometry inequalities and basic algebra inequality problems. Here is another example:

[Ex16.11] (International Olympics Mathematics Competition, 1984) A point P^* is in the interior of $\triangle ABC$. Line AP^*, BP^* and CP^* intersect BC, CA and AB at point P, Q and R, respectively. What position of point P^* makes the area of $\triangle PQR$ the biggest, as shown in Fig.16.14?

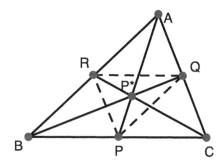

FIGURE 16.14

Solution: The key is to find the relationship between the position of P^* and the area of $\triangle PQR$.

Let

$$\frac{AQ}{QC} = \lambda, \qquad \frac{AR}{RB} = \mu$$

When λ and μ are determined, the position of P^* is defined. Then:

$$\frac{BP}{PC} = \frac{\triangle BP^*A}{\triangle AP^*C} = \frac{\triangle BP^*A}{\triangle BP^*C} \cdot \frac{\triangle BP^*C}{\triangle AP^*C}$$

$$= \frac{AQ}{QC} \cdot \frac{RB}{AR} = \frac{\lambda}{\mu}$$

$$\therefore \frac{\triangle ARQ}{\triangle ABC} = \frac{AR}{RB} \cdot \frac{AQ}{AC} = \frac{\lambda}{(1+\lambda)} \cdot \frac{\mu}{(1+\mu)}$$

$$\frac{\triangle BPR}{\triangle ABC} = \frac{BP}{BC} \cdot \frac{BR}{AB} = \frac{\lambda}{(1+\lambda)} \cdot \frac{\mu}{(1+\mu)}$$

$$\frac{\triangle CPQ}{\triangle ABC} = \frac{PC}{BC} \cdot \frac{CQ}{AC} = \frac{\lambda}{(1+\lambda)} \cdot \frac{\mu}{(1+\mu)}$$

$$\therefore \frac{\triangle PQR}{\triangle ABC} = 1 - \frac{\lambda\mu}{(1+\lambda)(1+\mu)} - \frac{\lambda}{(1+\lambda)(1+\mu)} - \frac{\mu}{(1+\lambda)(1+\mu)}$$

$$= \frac{2\lambda\mu}{(1+\lambda)(1+\mu)(\lambda+\mu)}$$

$$= \frac{2\lambda\mu}{2\lambda\mu + \lambda^2 + \mu^2 + \mu(1+\lambda^2) + \lambda(1+\mu^2)} \leq \frac{1}{4}$$

This is because $\lambda^2 + \mu^2 \geq 2\lambda\mu$, $1 + \lambda^2 \geq 2\lambda$, $1 + \mu^2 \geq 2\mu$. If and only if $\lambda = \mu = 1$, the equal sign in the inequalities above applies. So, when P^* is the centroid of $\triangle ABC$, the area of $\triangle PQR$ is biggest and the area is $\frac{1}{4}$ of $\triangle ABC$.

[Ex16.12] (National Mathematics League Competition, China, 1990)

Given: in $\triangle ABC$, $AB = AC$. Take a point P from the interior of $\triangle ABC$ and line AP intersects BC at point N.

Let

$$\angle NPC = \angle BAC = 2\angle BPN$$

As shown in Fig.16.15.

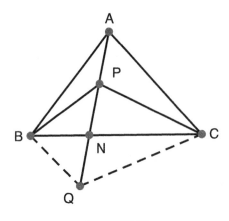

FIGURE 16.15

Prove:

$$BN = \frac{1}{3}BC$$

Solution: Take any point Q on line AN, we have

$$\frac{\triangle QPC}{\triangle QPB} = \frac{NC}{BN}$$

We want to prove the ratio of the equation is 2. But, what position of point Q can make it more convenient to compare the area of the two triangles?

Having noticed that $\angle QPC = 2\angle QPB$, you will easily find that if $QP = PC$, we have an isosceles triangle $\triangle PQC$. Then $\triangle PQC$ is divided into two parts, and one of the two parts is congruent with $\triangle PQB$. Follow this clue, and it is very easy to find the proof below:

Extend PN to point Q such that $PQ = PC$.

From $\angle QPC = \angle BAC$, we know $\angle PQC = \angle ABC$, thus A, B, C and Q lie on a circle.

Then $\angle PQB = \angle ACB = \angle PQC$ can be deduced. So $\triangle PQB$ is one half of $\triangle PQC$. Then $CN = 2BN$.

CHAPTER 17

INSTRUCTIONS, TIPS, AND SOLUTIONS TO PROBLEMS

This chapter gives the solutions and discussions to the "Additional Problems" section in chapters 1 through 15.

CHAPTER ONE

[P1.1]

$$\frac{\triangle PAB}{\triangle QAB} = \frac{PM}{QM}$$

[P1.2]

$$\frac{\triangle AOB}{\triangle AOC} = \frac{\triangle AOB}{\triangle AOP} \cdot \frac{\triangle AOP}{\triangle POC} \cdot \frac{\triangle POC}{\triangle BOC} \cdot \frac{\triangle BOC}{\triangle BOQ} \cdot \frac{\triangle BOQ}{\triangle AOQ} \cdot \frac{\triangle AOQ}{\triangle AOC}$$

$$= \frac{BO}{PO} \cdot \frac{AP}{PC} \cdot \frac{PO}{BO} \cdot \frac{CO}{QO} \cdot \frac{BQ}{AQ} \cdot \frac{QO}{CO}$$

$$= \frac{AP}{PC} \cdot \frac{BQ}{AQ} = \frac{4}{3} \cdot \frac{2}{3} = \frac{8}{9}$$

CHAPTER TWO

[P2.1]

$$\frac{PR}{BR} = \frac{\triangle PQC}{\triangle BQC} = \frac{\triangle PQC}{\triangle AQC} \cdot \frac{\triangle AQC}{\triangle BQC}$$

$$= \frac{PC}{AC} \cdot \frac{AQ}{BQ} = \frac{1}{2} \cdot \frac{2}{3} = 1$$

$$\frac{QR}{CR} = \frac{\triangle PQB}{\triangle PCB} = \frac{\triangle PQB}{\triangle AQC} \cdot \frac{\triangle PBA}{\triangle PCB}$$

$$= \frac{BQ}{AB} \cdot \frac{PA}{PC} = \frac{1}{3} \cdot \frac{1}{1} = \frac{1}{3}$$

$$\frac{\triangle RBC}{\triangle ABC} = \frac{\triangle RBC}{\triangle PBC} \cdot \frac{\triangle PBC}{\triangle ABC}$$

$$= \frac{BR}{BP} \cdot \frac{PC}{AC} = \frac{1}{2} \cdot \frac{1}{2} = \frac{1}{4}$$

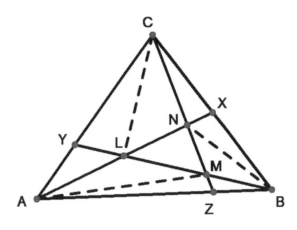

FIGURE 17.1

[P2.2] As shown in Fig. 17.1.

$$\frac{\triangle ABC}{\triangle ABL} = \frac{\triangle ABL + \triangle ACL + \triangle BCL}{\triangle ABL}$$

$$= 1 + \frac{CX}{BX} + \frac{CY}{AY}$$

$$= 1 + \frac{1}{1} + \frac{2}{1} = 4$$

$$\therefore \quad \triangle ABL = \frac{1}{4} \triangle ABC$$

$$\frac{\triangle ABC}{\triangle BCM} = \frac{\triangle BCM + \triangle ACM + \triangle ABM}{\triangle BCM}$$

$$= 1 + \frac{AZ}{BZ} + \frac{AY}{CY}$$

$$= 1 + 3 + \frac{1}{2} = \frac{9}{2}$$

$$\therefore \quad \triangle BCM = \frac{2}{9} \triangle ABC$$

$$\frac{\triangle ABC}{\triangle ACN} = \frac{\triangle ACN + \triangle BCN + \triangle ABN}{\triangle ACN}$$

$$= 1 + \frac{BZ}{AZ} + \frac{BX}{CX}$$

$$= 1 + \frac{1}{3} + \frac{1}{1} = \frac{7}{3}$$

$$\therefore \triangle ACN = \frac{3}{7} \triangle ABC$$

$$\therefore \triangle LMN = \triangle ABC - \triangle ABL - \triangle BCM - \triangle ACN$$

$$= \left(1 - \frac{1}{4} - \frac{2}{9} - \frac{3}{7}\right) \triangle ABC = \frac{25}{252} \triangle ABC$$

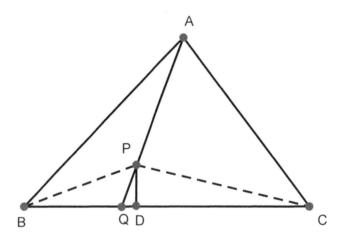

FIGURE 17.2

[P2.3] As shown in Fig. 17.2, we may suppose that $\triangle PCB$ is the smallest one of $\triangle PBC$, $\triangle PCA$, $\triangle PAB$. Then $\triangle PBC \leq \frac{1}{3} \triangle ABC$. Connect AP, and AP intersects BC at Q. Draw PD perpendicular to BC, because

$$\frac{PQ}{AQ} = \frac{\triangle PBC}{\triangle ABC} \leq \frac{1}{3}$$

$$\therefore \quad PD \leq PQ \leq \frac{1}{2} PA$$

[P2.4] Hint: Using the equation

$$\triangle ABC = \triangle PBC + \triangle PAB - \triangle PAC$$

192

$$\frac{PX}{AX} = \frac{\triangle PBC}{\triangle ABC}, \quad \frac{PZ}{CZ} = \frac{\triangle PAB}{\triangle ABC}, \quad \frac{PY}{BY} = \frac{\triangle PAC}{\triangle ABC}$$

[P2.5]

$$\frac{AX}{XB} \cdot \frac{BZ}{ZC} \cdot \frac{CY}{YA} = \frac{\triangle AXZ}{\triangle BXZ} \cdot \frac{\triangle BXZ}{\triangle CXZ} \cdot \frac{\triangle CXZ}{\triangle AXZ} = 1$$

[P2.6]

$$\frac{DN}{CN} = \frac{\triangle MDN}{\triangle MCN} = \frac{\triangle MDN}{\triangle MAN} \cdot \frac{\triangle MAN}{\triangle MBN} \cdot \frac{\triangle MBN}{\triangle MCN}$$

$$= \frac{PD}{AP} \cdot \frac{AM}{BM} \cdot \frac{BQ}{CQ} = \frac{AM}{BM}$$

Notice: from

$$\frac{PD}{AD} = \frac{QC}{BC} \quad \text{we have} \quad \frac{PD}{AP} = \frac{CQ}{BQ}$$

[P2.7]

$$\frac{PY}{PX} = \frac{\triangle APY}{\triangle APX} > \frac{\triangle APB}{\triangle APC} = \frac{PB}{PC}$$

CHAPTER THREE

[P3.1]

$$\frac{PO}{QO} = \frac{\triangle APC}{\triangle AQC} = \frac{2\triangle APO}{2\triangle COQ} = \frac{AO}{CO} = 1$$

[P3.2] It's sufficient to prove $\triangle PBC = \triangle QBC$.

$$\frac{\triangle PBC}{\triangle QBC} = \frac{\triangle PBC}{\triangle ABC} \cdot \frac{\triangle ABC}{\triangle QBC} = \frac{PA}{AB} \cdot \frac{AC}{QC}$$

Notice: from the condition we have

$$\frac{PA}{BA} = \frac{QA}{CA} \quad \text{we have} \quad \frac{PB}{AB} = \frac{QC}{AC}$$

[P3.3] It's sufficient to prove $\triangle PMO = \triangle PNO$.

$$\frac{\triangle PMO}{\triangle PNO} = \frac{\triangle PMO}{\triangle MNO} \cdot \frac{\triangle MNO}{\triangle PNO} = \frac{PB}{NB} \cdot \frac{MA}{PA}$$

$$= \frac{\triangle ABP}{\triangle ABN} \cdot \frac{\triangle ABM}{\triangle ABP} = 1$$

[P3.4]

$$\frac{MG}{NH} = \frac{MG}{MH} \cdot \frac{MH}{NH} = \frac{\triangle ADG}{\triangle ADH} \cdot \frac{\triangle MBD}{\triangle BND}$$

And because

$$\frac{\triangle ADG}{\triangle BND} = \frac{\triangle ADG}{\triangle DGC} \cdot \frac{\triangle DGC}{\triangle DNC} \cdot \frac{\triangle DNC}{\triangle BND}$$

$$= \frac{AG}{GC} \cdot \frac{1}{1} \cdot \frac{CN}{BN}$$

$$= \frac{\triangle AMN}{\triangle CMN} \cdot \frac{\triangle CMN}{\triangle BMN}$$

$$= \frac{\triangle AMN}{\triangle BMN} = 1$$

$$\frac{\triangle MBD}{\triangle ADH} = \frac{\triangle MDH + \triangle MBH}{\triangle MDH + \triangle MAH} = 1$$

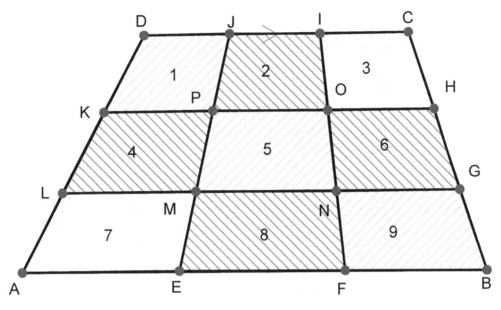

CHAPTER FOUR

FIGURE 17.3

[P4.1] Hint: First prove:

$$S_{AEJD} = S_{EFIJ} = S_{FBCI} = \frac{S}{3}$$

Therefore it is sufficient to calculate the area of 1,4,7.

According to Ex4.1, we know that $S_4 = \frac{1}{3}S_{AEJD} = \frac{S}{9}$, thus calculating S_1 and S_7 is necessary.

Based on the given $CD = \frac{2}{3}AB$, we have $DJ = \frac{2}{3}AE$

$$\therefore \frac{\triangle DJA}{\triangle AJE} = \frac{DJ}{AE} = \frac{2}{3}$$

$$\therefore \triangle DJA = \frac{2}{15}S, \quad \triangle AJE = \frac{3}{15}S$$

$$\therefore S_1 = S_{KPJD} = \triangle DKJ + \triangle KPJ$$

$$= \frac{1}{3}\triangle DAJ + \left(\frac{1}{3}\triangle KEJ\right)$$

$$= \frac{2}{45}S + \frac{1}{3}\left(\frac{2}{3}\triangle DJE + \frac{1}{3}\triangle AJE\right)$$

$$= \frac{2}{45}S + \frac{1}{3}\left(\frac{2}{3}\triangle DJA + \frac{3}{15}S\cdot\frac{1}{3}\right)$$

$$= \frac{2}{45}S + \frac{1}{3}\left(\frac{2}{3}\cdot\frac{2}{15}S + \frac{S}{15}\right) = \frac{13}{135}S$$

$$\therefore\ S_7 = \frac{S}{3} - \frac{S}{9} - \frac{13}{135}S = \frac{17}{135}S$$

[P4.2] If it is divided equally into 4 parts, then the average of the 4 middle parts is $\frac{1}{16}$ of S_{ABCD}. If it is divided equally into 5 parts, then the middle one is $\frac{1}{25}$ of S_{ABCD}. Proof is similar to the trisection case. The key is to make flexible use of the Point of Division Formula.

[P4.3] If we know the area of the $\triangle ABD$, $\triangle ABC$, $\triangle BCD$ from

$$\triangle ABD + \triangle ACD = \triangle ABD + \triangle BCD,$$

then we know the area of $\triangle ACD$. Using the Point of Division Formula, we have

$$\begin{cases} \triangle DJE = \dfrac{1}{3}\triangle DCE = \dfrac{1}{3}\left(\dfrac{2}{3}\triangle ACD + \dfrac{1}{3}\triangle BCD\right) \\[2mm] \triangle AJE = \dfrac{1}{3}\triangle ABJ = \dfrac{1}{3}\left(\dfrac{2}{3}\triangle ABD + \dfrac{1}{3}\triangle ABC\right) \\[2mm] \triangle ADE = \dfrac{1}{3}\triangle ABD \\[2mm] \triangle ADJ = \dfrac{1}{3}\triangle ACD \end{cases}$$

Then

$$S_{AEJD} = \triangle ADE + \triangle DJE$$

$$= \frac{1}{9}\triangle BCD + \frac{2}{9}\triangle ACD + \frac{3}{9}\triangle ABD$$

And because *K and L* are the trisection points of *AD and P and M* are the trisection points of *JE*, the area of three quadrilaterals making up *AEJD* can be calculated. The area of the other pieces can also be calculated by using the similar methods.

Or directly use the following formula

$$\triangle DKJ = \frac{1}{3} \triangle ADJ$$

$$\triangle PKJ = \frac{1}{3} \triangle EKJ = \frac{1}{3}\left(\frac{1}{3} \triangle EAJ + \frac{2}{3} \triangle EDJ\right)$$

CHAPTER FIVE

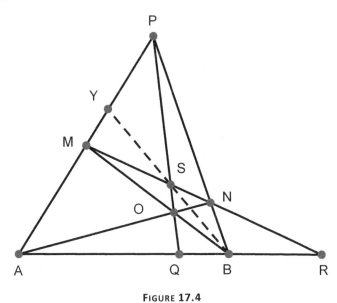

FIGURE 17.4

[P5.1] The following proportions exist:

$$\frac{BX}{SX} = \frac{BY}{SY}, \quad \frac{PY}{MY} = \frac{PA}{MA} \quad \text{and} \quad \frac{NX}{OX} = \frac{NA}{OA}$$

The proof is to correspond the points *A, B, P, O, M, N, S, Q, R* in the Ex15.1 with the points *P, M, B, S, N, O, X, Y, A* in the Fig. 17.4. Then prove the ratios using the method of **[Ex5.1]**.

[P5.2]

$$\frac{AQ}{BQ} = \frac{\triangle AOP}{\triangle BOP} = \frac{\triangle AOP}{\triangle PON} \cdot \frac{\triangle PON}{\triangle BOP}$$

$$= \frac{AO}{NO} \cdot \frac{PN}{PB}$$

$$= \frac{\triangle ABM}{\triangle NBM} \cdot \frac{\triangle ANM}{\triangle ABM}$$

$$= \frac{\triangle ANM}{\triangle NBM} = \frac{AR}{BR}$$

[P5.3] The order of drawing steps in Ex2.4 is: first take the free points A, B, C, P such that AP intersects BC at X, BP intersects AC at Y, CP intersects AB at Z. Then the process of eliminating points is :

Eliminating Z from $\quad \dfrac{PZ}{CZ}$, $\quad$ we have $\quad \dfrac{\triangle PAB}{\triangle ABC}$.

The order of drawing steps in Ex2.7 is: first take the free points A, B, C, and take D such that $\quad AD = BC$, and take the midpoint M of AB, the midpoint N of CD, the intersection point P of MN and AD, the intersection point Q of MN and BC.

Then the process of eliminating points is:

Start from

$$\frac{PD}{AD} \cdot \frac{BC}{QC}$$

$$\frac{PD}{AD} \cdot \frac{BC}{QC} = \frac{PD}{AD} \cdot \left(\frac{\triangle BMN - \triangle CMN}{\triangle CMN} \right) (Q \ is \ eliminated)$$

$$= \frac{\triangle DMN}{(\triangle AMN - \triangle DMN)} \cdot \left(\frac{\triangle BMN - \triangle CMN}{\triangle CMN} \right) (P \ is \ eliminated)$$

$$= \left(\frac{\triangle DMN}{\triangle CMN}\right) \cdot \left(\frac{\triangle BMN - \triangle CMN}{\triangle AMN - \triangle DMN}\right) = 1$$

We use $\triangle DMN = \triangle CMN$, $\triangle AMN = \triangle BMN$ to eliminate points M and N.

The order of drawing steps in Ex3.5 is: first take the free points A, B, P. Take a point M on AP, and draw a parallel of AB through M such that the line intersects PB at N. AN intersects BM at O, and PO intersects AB at Q.

Start from $\dfrac{AQ}{BQ}$ to eliminate points:

$$\frac{AQ}{BQ} = \frac{\triangle AOP}{\triangle BOP} \, (Q \ is \ eliminated)$$

$$= \frac{\triangle AOP}{\triangle ABP} \cdot \frac{\triangle ABP}{\triangle BOP}$$

$$= \frac{MO}{MB} \cdot \frac{AN}{NO}$$

$$= \frac{\triangle AMN}{S_{ABNM}} \cdot \frac{S_{ABNM}}{\triangle BMN} \, (O \ is \ eliminated)$$

$$= \frac{\triangle AMN}{\triangle BMN} = 1 (M \ and \ N \ are \ eliminated)$$

CHAPTER SIX

[P6.1] The drawing order is:

(1) Free points A, B, E, D.

(2) Half free point: Take any point C on AB.

(3) Constraint points:

J—the intersection point of AE and CD,

I—the intersection point of AE and BD,

H—the intersection point of *AD* and *CE,*

G—the intersection point of *AD* and *BE.*

The conclusions are transformed as follows:

For proving *IH, JG, AB* have the same intersection point, we may prove that the intersection point of *IH* and *AB* overlaps with the intersection point of *JG* and *AB*. So we draw auxiliary points:

(4) auxiliary constraint points:

O—the intersection point of *JG* and *AB*

P—the intersection point of *IH* and *AB*

To prove *O* is coincident with *P*, we may prove

$$\frac{AO}{BO} \cdot \frac{BP}{AP} = 1$$

Then the process of eliminating points is:

(1) $\dfrac{AO}{BO} \cdot \dfrac{BP}{AP} = \dfrac{\triangle AJG}{\triangle BJG} \cdot \dfrac{\triangle BIH}{\triangle AIH}$ (*O* and *P* are eliminated)

(2)

$$\left.\begin{array}{l}
\triangle AJG = \dfrac{\triangle AJG}{\triangle AJD} \cdot \triangle AJD = \dfrac{AG}{AD} \cdot \triangle AJD = \dfrac{\triangle ABE}{S_{ABDE}} \cdot \triangle AJD \\[3mm]
\triangle BJG = \dfrac{\triangle BJG}{\triangle BJE} \cdot \triangle BJE = \dfrac{BG}{BE} \cdot \triangle BJE = \dfrac{\triangle ABD}{S_{ABDE}} \cdot \triangle BJE
\end{array}\right\} \text{(}G \text{ is eliminated)}$$

$$\left.\begin{array}{l}
\triangle BIH = \dfrac{\triangle BIH}{\triangle BIA} \cdot \triangle BIA = \dfrac{HD}{AD} \cdot \triangle BIA = \dfrac{\triangle CDE}{S_{ACDE}} \cdot \triangle BIA \\[3mm]
\triangle AIH = \dfrac{\triangle AIH}{\triangle AID} \cdot \triangle AID = \dfrac{AH}{AD} \cdot \triangle AID = \dfrac{\triangle ACE}{S_{ACDE}} \cdot \triangle AID
\end{array}\right\} \text{(}H \text{ is eliminated)}$$

Plugging them into the equation above, we have

$$\frac{\triangle AJG}{\triangle BJG} \cdot \frac{\triangle BIH}{\triangle AIH} = \frac{\triangle ABE}{\triangle ABD} \cdot \frac{\triangle CDE}{\triangle ABD} \cdot \frac{\triangle AJD}{\triangle BJE} \cdot \frac{\triangle BIA}{\triangle AID}$$

(3)

$$\triangle AJD = \frac{\triangle AJD}{\triangle ADC} \cdot \triangle ADC = \frac{JD}{DC} \cdot \triangle ADC = \frac{\triangle ADE \cdot \triangle ADC}{\triangle ACE - \triangle ADE} \Bigg\}$$

$$\triangle BJE = \frac{\triangle BJE}{\triangle BEA} \cdot \triangle BEA = \frac{JE}{EA} \cdot \triangle BEA = \frac{\triangle CDE \cdot \triangle ABE}{\triangle ACD - \triangle ECD} \Bigg\} \quad (K \text{ is eliminated})$$

$$\triangle BIA = \frac{\triangle BIA}{\triangle BDA} \cdot \triangle BDA = \frac{BI}{BD} \cdot \triangle BDA = \frac{\triangle ABE}{S_{ABED}} \cdot \triangle ABD \Bigg\}$$

$$\triangle BIA = \frac{\triangle AID}{\triangle ADE} \cdot \triangle ADE = \frac{AI}{EA} \cdot \triangle ADE = \frac{\triangle ABD}{S_{ABED}} \cdot \triangle ADE \Bigg\} \quad (J \text{ is eliminated})$$

Plugging them into the equation above, we have

$$\frac{\triangle ABE}{\triangle ABD} \cdot \frac{\triangle CDE}{\triangle ACE} \cdot \frac{\triangle AJD}{\triangle BJE} \cdot \frac{\triangle BIA}{\triangle AID}$$

$$= \frac{\triangle ABE}{\triangle ABD} \cdot \frac{\triangle CDE}{\triangle ACE} \cdot \frac{\triangle ADE \cdot \triangle ADC}{\triangle CDE \cdot \triangle ABE} \cdot \frac{\triangle ABE \cdot \triangle ABD}{\triangle ABD \cdot \triangle ADE}$$

$$= \frac{\triangle ADC \cdot \triangle ABE}{\triangle ACE \cdot \triangle ABD}$$

$$= \frac{\triangle ADC}{\triangle ABD} \cdot \frac{\triangle ABE}{\triangle ACE}$$

$$= \frac{AC}{AB} \cdot \frac{AB}{AC} = 1$$

CHAPTER SEVEN

[P7.1] Notice that $\angle BAE + \angle CAE = 180°$, so we have

$$\frac{BE}{CE} = \frac{\triangle ABE}{\triangle ACE} = \frac{AB \cdot AE}{AC \cdot AE} = \frac{AB}{AC}$$

[P7.2] Suppose M and N are the midpoints of AB and AC respectively, then from

$\triangle MBC = \frac{1}{2} \triangle ABC = \triangle NBC$, we know that $MN \parallel BC$. Thus $\angle AMN = \angle ABC$. Based on Co-Angle Theorem, we have

$$\frac{AM \cdot AN}{AB \cdot AC} = \frac{\triangle AMN}{\triangle ABC} = \frac{AM \cdot MN}{AB \cdot BC}$$

$$\therefore \quad \frac{MN}{BC} = \frac{AN}{AC} = \frac{1}{2}$$

[P7.3] (1) Use $ab = 2\triangle ABC = ch$

(2) Use $\triangle ACD + \triangle BCD = \triangle ABC$;

(3) Associate (1) with (2).

[P7.4] Hint:

$$\frac{BN}{BD} = \frac{\triangle BPQ}{S_{BPDQ}} = \frac{\triangle PBQ}{\triangle ABC} \cdot \triangle ABC \cdot \frac{1}{\triangle PBD + \triangle BQD}$$

But

$$\frac{\triangle PBQ}{\triangle ABC} = \frac{PB}{AB} \cdot \frac{BQ}{BC} = \frac{3}{4} \cdot \frac{2}{3} = \frac{1}{2}$$

$$\triangle PBD = \frac{3}{4} \triangle ABC, \qquad \triangle BQD = \frac{2}{3} \triangle ABC$$

[P7.5] Hint:

$$\frac{AP}{PB} = \frac{\triangle ACP}{\triangle PCB} = \frac{\triangle ACP}{\triangle AMC} = \frac{\triangle MAC}{\triangle PCB}$$

And

$$\angle ACP = \angle AMC, \quad \angle PCB = \angle MAC.$$

[P7.6] Hint:

$$\frac{AM}{AN} = \frac{\triangle AEM + \triangle AMF}{\triangle AEF}$$

$$\triangle AEF = \frac{AE \cdot AF}{AB \cdot AC} \cdot \triangle ABC$$

$$\triangle AEM = \frac{AE}{AB} \triangle ABM$$

$$\triangle ABM = \frac{BM}{BC} \triangle ABC$$

Similarly, we can calculate $\triangle AMF$.

$$\frac{AM}{AN} = \frac{AM}{AM - NM} = \frac{1}{1 - \dfrac{NM}{AM}}$$

[P7.7] Hint:

$$\frac{NM}{AM} = \frac{\triangle BNC}{\triangle ABC}$$

$$\triangle BNC = \frac{\triangle BEC + \triangle BFC}{2}$$

[P7.8]

$$\begin{aligned}
\frac{MG}{AG} \cdot \frac{BH}{NH} &= \frac{\triangle MCF}{\triangle ACF} \cdot \frac{\triangle BDE}{\triangle NDE} \\
&= \frac{BD \cdot BE}{AF \cdot AC} \cdot \frac{MF \cdot OC}{OE \cdot ND} \\
&= \frac{BD \cdot BE}{AF \cdot AC} \cdot \frac{\triangle BMF}{\triangle BOE} \cdot \frac{\triangle AOC}{\triangle AND} \\
&= \frac{BD \cdot BE}{AF \cdot AC} \cdot \frac{BM \cdot AF}{AN \cdot BD} \cdot \frac{AC \cdot QO}{BE \cdot QO} \\
&= \frac{BM}{AN}
\end{aligned}$$

CHAPTER EIGHT

[P8.1] Extend AB to P, $A'B'$ to P' such that $AB = BP, A'B' = B'P'$,then $\triangle ABC = \triangle PBC$, $\triangle A'B'C' = \triangle P'B'C'$ and $\angle ABC = 180° - \angle PBC$, $\angle A'B'C' = 180° - \angle P'B'C'$, Thus $\angle PBC + \angle P'B'C' < 180°$ and $\angle P'B'C' > \angle PBC$. Based on the Co-Angle Theorem, we have

$$\begin{aligned}
\frac{\triangle A'B'C'}{\triangle ABC} &= \frac{\triangle P'B'C'}{\triangle PBC} > \frac{P'B' \cdot B'C'}{PB \cdot BC} \\
&= \frac{A'B' \cdot B'C'}{AB \cdot BC}
\end{aligned}$$

[P8.2] This is not always the case. For example, suppose $\triangle ABC$ is an equilateral triangle and $\triangle A'B'C'$ is an isosceles right triangle. And suppose $AB = BC = A'B' = A'C' = 1$, then $\triangle ABC = \frac{\sqrt{3}}{4}$ and $\triangle A'B'C' = \frac{1}{2}$.

[P8.3] Hint: Rotating a triangle along the midpoint of one side to $180°$, we get a parallelogram. Then the median becomes $\frac{1}{2}$ of one diagonal line of the parallelogram.

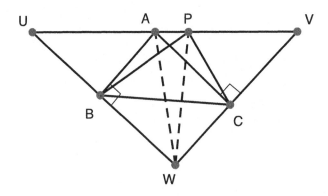

FIGURE 17.5

[P8.4] As shown in Fig.17.5, suppose $AP \parallel BC$, then draw perpendiculars AB and AC to UW and VW forming the isosceles triangle $\triangle WUV$ where UV passes through A and P. Suppose $l = WU = WV$, then we have

$$
\begin{aligned}
l(AB + AC) &= 2 \triangle AWU + 2 \triangle AWV \\
&= 2 \triangle WUV \\
&= 2 \triangle PWU + 2 \triangle PWV \\
&< l(PB + PC)
\end{aligned}
$$

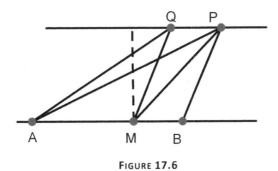

FIGURE 17.6

[P8.5] As shown in Fig.17.6, if we can prove $\angle AQB > \angle APB$, then we have the conclusion immediately. This is equivalent to $\angle QAP < \angle QBP$. Refer to the proof of Ex8.7.

[P8.6] Refer to the proof of Ex8.8.

CHAPTER NINE

[P9.1] Since $\angle 3$ and $\angle 4$ in the proof may be $0°$ (i.e. A, B, A' may be collinear), then we can not get $AB = A'B'$.

[P9.2] This exercise is a deformation of Ex9.5. We can use a similar proof:

$$\frac{AQ}{DQ} = \frac{\triangle AMN}{\triangle DMN} = \frac{\triangle BMN}{\triangle CMN} = \frac{BP}{CP}$$

(because M and N are the midpoints of AB and CD respectively)

That is

$$\frac{AQ}{AD - AQ} = \frac{BP}{BC - BP}$$

From $AD=BD=BC$, we get $AQ=BP$.

Extend PM to P' such that $PM = P'M$. Then we know easily that $\triangle AMP' \cong \triangle BMP$, thus $AP' = BP = AQ$, so

$$\angle MP'A = \angle AQM$$

But

$$\angle MP'A = \angle BPM,$$

thus $\angle BPM = \angle AQM$, which is equivalent to $\angle BPN = \angle DQN$.

CHAPTER TEN

[10.1] Hint: From the area formula of triangle $S = \frac{1}{2}ah$, we can get the conclusion.

[10.2] In $\triangle ABC$, let h, x, y be the altitude from C, the perpendicular distance from P to side AB and the perpendicular distance from P to side AC, respectively. Let $AB = AC = b$. From $\triangle APC + \triangle APB = \triangle ABC$, we have

$$\frac{1}{2}bx + \frac{1}{2}by = \frac{1}{2}hb$$

[10.3] Suppose the distance from G to BC, CA, AB is x, y, z respectively, and then we get

$$\triangle GBC = \frac{1}{2}x \cdot BC, \triangle GCA = \frac{1}{2}y \cdot CA$$

$$\triangle GAB = \frac{1}{2}z \cdot AB$$

Because G is the centroid of a triangle, we can prove that

$\triangle GBC = \triangle GCA = \triangle GAB$. From this, one may reach the desired conclusion.

[10.4] Use Area Equation $\triangle BCN = \triangle BCM + \triangle MCN$, divide both sides by $\triangle BDC$, and we have:

$$\frac{\triangle BCN}{\triangle BDC} = \frac{CN}{DC} = k$$

$$\frac{\triangle BCM}{\triangle BDC} = \frac{\triangle BCM}{\triangle BCA} \cdot \frac{\triangle BCA}{\triangle DCA} \cdot \frac{\triangle DCA}{\triangle DCO} \cdot \frac{\triangle DCO}{\triangle BDC}$$

$$= \frac{CM}{CA} \cdot \frac{BO}{DO} \cdot \frac{CA}{CO} \cdot \frac{DO}{BD}$$

$$= (1 - k) \cdot 2 \cdot \frac{BO}{BD}$$

$$\frac{\triangle MCN}{\triangle BDC} = \frac{\triangle MCN}{\triangle DCA} \cdot \frac{\triangle DCA}{\triangle DCO} \cdot \frac{\triangle DCO}{\triangle BDC}$$

$$= \frac{MC \cdot NC}{AC \cdot CD} \cdot 2 \cdot \frac{DO}{BD}$$

$$= 2k(1-k) \cdot \frac{DO}{BD}$$

Suppose
$$\frac{DO}{BO} = x$$

and then we have
$$\frac{BO}{BD} = \frac{1}{1+x}, \qquad \frac{DO}{BD} = \frac{x}{1+x}$$

Plugging them into the Area Equation, we get:

$$k = 2(1-k)\frac{1}{1+x} + 2k(1-k)\frac{x}{1+x}$$

By rearrangement: $\quad k(1+x) = 2(1-k) + 2k(1-k)x$

CHAPTER ELEVEN

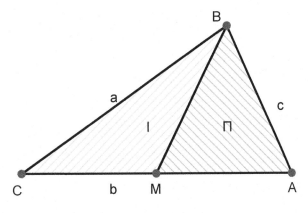

FIGURE 17.7

[P11.1] From Pythagorean Difference Theorem, as shown in the Fig. 17.7 , we have

$$\frac{b^2 + a^2 - c^2}{\left(\frac{b}{2}\right)^2 + a^2 - l^2} = \frac{\triangle ABC}{\triangle I} = 2$$

Then we can get the solution:

[P11.2] If $a^2 + b^2 - c^2 = 0$, then $a'^2 + b'^2 - c'^2 = 0$, then $\angle C = \angle C' = 90°$. In the following, we suppose $a^2 + b^2 - c^2 \neq 0$, then $a'^2 + b'^2 - c'^2 \neq 0$. Notice that

$$\frac{\triangle ABC}{\triangle A'B'C'} > 0,$$ so both $\angle C$ and $\angle C'$ are acute angles or obtuse angles.

If so, we just prove the case: $\angle C$ and $\angle C'$ are acute angles. Suppose $\angle C < \angle C'$. Construct another triangle $\triangle A_1B_1C_1$, such that $A_1C_1 = b'$, $B_1C_1 = a'$ and $\angle A_1B_1C_1 = \angle C$. Thus, we get

$$A_1B_1 < c', \qquad \triangle A_1B_1C_1 < \triangle A'B'C'$$

Let $a_1 = B_1C_1 = a'$, $b_1 = A_1C_1 = b'$, $c_1 = A_1B_1$

Then

$$\therefore \frac{a^2 + b^2 - c^2}{a_1{}^2 + b_1{}^2 - c_1{}^2} < \frac{a^2 + b^2 - c^2}{a'^2 + b'^2 - c'^2}$$

$$= \frac{\triangle ABC}{\triangle A'B'C'} < \frac{\triangle ABC}{\triangle A_1B_1C_1}$$

It contradicts the Pythagorean Difference Theorem.

[P11.3] It's appropriate as long as we prove: when a and b are invariant, $\angle C$ is bigger, and the Pythagorean Difference of $\angle C'$ in $\triangle ABC$ is smaller. Then, just take into account the two cases: acute angle and obtuse angle.

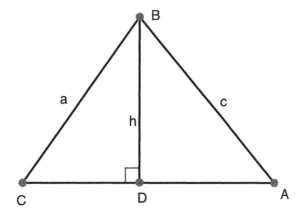

FIGURE **17.8**

[P11.4] If $\angle C$ is a right angle then $\quad \triangle ABC = \dfrac{ab}{2}$, $\quad$ obviously. In the following, we prove the case where $\angle C < 90°$.

As shown in Fig.17.8, suppose $h = BD$, which is an altitude of $\triangle ABC$. From Pythagorean Difference Theorem, we have:

$$\frac{a^2 + b^2 - c^2}{a^2 + CD^2 - h^2} = \frac{\triangle ABC}{\triangle DBC} = \frac{AC}{DC}$$

Suppose $\frac{AC}{DC} = k$, and we have

$$DC = \frac{b}{k}, \qquad DBC = \frac{\triangle ABC}{k}$$

$$a^2 + CD^2 - h^2 = \frac{a^2 + b^2 - c^2}{k}\left(\frac{a^2 + CD^2 - h^2}{2ab}\right) + \left(\frac{2 \triangle DBC}{ab}\right)^2$$

$$= \left(\frac{2CD}{2ab}\right)^2 + \left(\frac{CD \cdot h}{ab}\right)^2$$

$$= \left(\frac{CD}{ka}\right)^2 + \left(\frac{h}{ka}\right)^2$$

$$= \frac{1}{k^2}\left(\frac{CD^2 + h^2}{a^2}\right) = \frac{1}{k^2}$$

Let $\triangle DBC = \frac{1}{K} \triangle ABC$, $a^2 + CD^2 - h^2 = \frac{1}{K}(a^2 + b^2 - c^2)$, Plugging them into the left-hand side of the equation above, we get the equation that is to prove. From this equation, we have

$$(\triangle ABC)^2 = \left(\frac{ab}{2}\right)^2 - \left(\frac{a^2 + b^2 - c^2}{4}\right)^2$$

This is the Three Slanting Quadrature Formula.

CHAPTER TWELVE

[P12.1] Suppose h is the height of side BC in the $\triangle ABC$. In Ex12.2, we have already proved

$$d = \frac{AB \cdot AC}{h}$$

In addition

$$h = \frac{2 \triangle ABC}{BC}$$

Plugging this into the equation of d, we will have

$$d = \frac{BC \cdot CA \cdot AB}{2 \cdot ABC}$$

[P12.2] With the help of the Co-Side Theorem and Co-Circle Theorem, we get

$$\frac{PA}{PB} = \frac{\triangle ADC}{\triangle BDC} = \frac{AD \cdot AC \cdot DC}{BD \cdot BC \cdot DC}$$
$$= \frac{AD \cdot AC}{BD \cdot BC}$$

[P12.3] Remark: When one diagonal line of the quadrilateral *ABCD* or *WXYZ* slides along the line that the diagonal line is on, the area of the quadrilateral remains the same (why?). From this, the proposition is converted into the Common Angle Theorem.

[P12.4]

$$\frac{MG}{AG} \cdot \frac{BH}{MH} = \frac{\triangle MCF}{\triangle ACF} \cdot \frac{\triangle BDE}{\triangle MDE}$$

$$= \frac{\triangle BDE}{\triangle ACF} \cdot \frac{MC \cdot MF}{MD \cdot ME}$$

$$= \frac{\triangle BDE}{\triangle ACF} \cdot \frac{\triangle ACB}{\triangle ADB} \cdot \frac{\triangle AFB}{\triangle AEB}$$

$$= \frac{BD \cdot DE \cdot BE}{AC \cdot CF \cdot AF} \cdot \frac{AC \cdot BC \cdot AB}{AD \cdot BD \cdot AB} \cdot \frac{AF \cdot FB \cdot AB}{AE \cdot EB \cdot AB}$$

$$= \frac{DE \cdot BC \cdot FB}{CF \cdot AE \cdot AD} = \frac{DE}{CF} \cdot \frac{BC}{AD} \cdot \frac{FB}{AE}$$

$$= \frac{ME}{MC} \cdot \frac{MC}{MA} \cdot \frac{MB}{ME} = \frac{MB}{MA}$$

This is equivalent to the equation that we were assigned to prove.

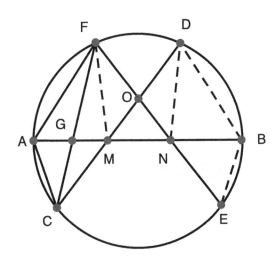

FIGURE 17.9

12.5 We will come to the same conclusion as exercise 7.8. As shown in Fig.17.9, we get:

$$\frac{MG}{AG} \cdot \frac{BH}{NH} = \frac{MB}{NA} \ .$$ The reasoning process is:

$$\frac{MG}{AG} \cdot \frac{BH}{NH} = \frac{\triangle MCF}{\triangle ACF} \cdot \frac{\triangle BDE}{\triangle NDE}$$

$$= \frac{\triangle BDE}{\triangle ACF} \cdot \frac{\triangle MCF}{\triangle NDE}$$

$$= \frac{\triangle BDE}{\triangle ACF} \cdot \frac{\triangle MCF}{\triangle DCF} \cdot \frac{\triangle DCF}{\triangle DEF} \cdot \frac{\triangle DEF}{\triangle DEN}$$

$$= \frac{BD \cdot DE \cdot BE}{AC \cdot AF \cdot CF} \cdot \frac{MC}{DC} \cdot \frac{DC \cdot CF}{DE \cdot EF} \cdot \frac{EF}{EN}$$

$$= \frac{BD \cdot BE \cdot MC}{AC \cdot AF \cdot EN}$$

$$= \frac{MB}{MC} \cdot \frac{NE}{NA} \cdot \frac{MC}{EN} = \frac{MB}{NA}$$

CHAPTER THIRTEEN

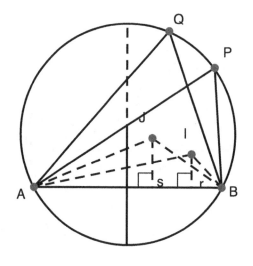

FIGURE 17.10

[P13.1] Hint: (1) Suppose the incenters of $\triangle PAB$ and $\triangle QAB$ are I and J, respectively, and then the heights of $\triangle IAB$ and $\triangle JAB$ are r and s which correspond to the radius of their inscribed circles.

(2) The desired conclusion is: $s > r$.

(3) Notice that $\angle AJB = \angle AIB$, so these four points A, J, I, B are on the same circle. Then we can get $s > r$.

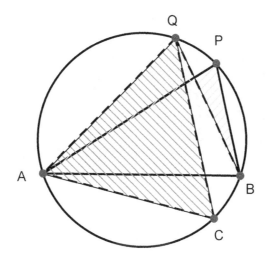

FIGURE 17.11

[P13.2] Hint: (1) The radius of the right triangle's circumcircle is twice that of its inscribed circle.

(2) If $\triangle PAB$ is not a right triangle, we suppose that the biggest angle is $\angle B$, the smallest angle is $\angle A$. Then we have $\angle B > 60° > \angle A$.

Take a point Q from PA such that $\angle QAB = 60°$. Then the inscribed circle of $\triangle QAB$ is bigger than the inscribed circle of $\triangle PAB$ (with the result of exercise 13.1).

(3) If $\triangle QAB$ is a right triangle, the problem is solved. If not, suppose $\angle QAB < \angle QBA$, and we take a point C on arc AB such that $\angle QBA = 60°$. Then the inscribed circle of $\triangle QAB$ is bigger than the inscribed circle of $\triangle QAC$. Here, $\triangle CQA$ is a right triangle. The problem has been dealt with.

[P13.3] Suppose O_1, O_2 are the midpoints of BE and DC respectively. From $BC \parallel DE$, we know $O_1O_2 \parallel BC \parallel DE$, so $\triangle DO_1O_2 = \triangle EO_1O_2$. Suppose O_1N and O_2M are the height of $\triangle DO_1O_2$ and $\triangle EO_1O_2$ respectively. Then we have $O_1N \cdot DO_2 = O_2M \cdot$

EO_1 When $\odot O_1$ is tangent with CD, we have $O_1N = O_1E$ and $O_2D = O_2M$. Thus, we deduce $\odot O_2$ is tangent to BE.

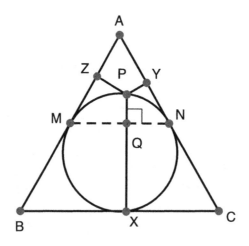

FIGURE 17.12

[P13.4] Hint: As shown in Fig.17.12, suppose M and N are the points of tangency, and P is on arc MN. Let PX intersect MN at Q. Since points M and N are the midpoints of AB and BC respectively, QX is half of the height of $\triangle ABC$.

Then $PY + PZ + PQ = XQ$.

Adding PQ on both sides, we get $PY + PZ + 2PQ = PX$

But the equality to prove is: $\quad PY + PZ + 2\sqrt{PY \cdot PZ} = PX$

Thus, it's solved as long as we have proved $\quad PY \cdot PZ = PQ^2$.

This is the result in Ex13.5.

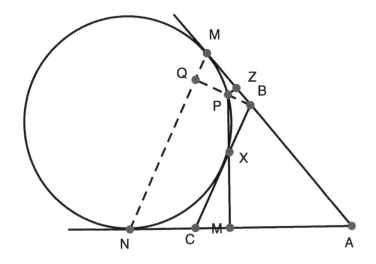

FIGURE 17.13

[P13.5] Hint: As shown in Fig.17.13, we have $PZ + PY - PX = h$ where h is the altitude of $\triangle ABC$. Suppose M and N are the points of tangency. It is easy to conclude $AB = 2BM$, $AC = 2CN$. Then we suppose PX intersects MN at Q, and we have

$$QX = \frac{1}{2}h = \frac{1}{2}(PZ + PY - PX)$$

$$PQ = \frac{3}{2}h - PZ - PY$$

$$\therefore \quad PX = QX - PQ = PZ + PY - h$$

$$= \frac{1}{3}(PY + PZ) - \frac{2}{3}PQ$$

$$\therefore \quad 3PX = PY + PZ - 2PQ$$

From Ex.13.5, $PQ = \sqrt{PY \cdot PZ}$

$$\therefore \quad \sqrt{3PX} = \left| \sqrt{PY} - \sqrt{PZ} \right|$$

CHAPTER FOURTEEN

[P14.1] When T is on the extension line PQ, Q is on the line segment PT. Since $PT = \lambda PQ$, then $PQ = \frac{1}{\lambda} PT$. Using the Point of Division Formula to Q, we get

$$\triangle QAB = \frac{1}{\lambda} \triangle TAB + \left(1 - \frac{1}{\lambda}\right) \triangle PAB$$

$$\therefore \quad \lambda \triangle QAB = \triangle TAB + (\lambda - 1) \triangle PAB$$

$$\therefore \quad \triangle TAB = \lambda \triangle QAB + (1 - \lambda) \triangle PAB$$

This shows that the original formula still holds here.

When the line segment PQ intersects line AB at M, and T is on the line segment QM, we have

$$\triangle TAB = \lambda \triangle QAB - (1 - \lambda) \triangle PAB$$

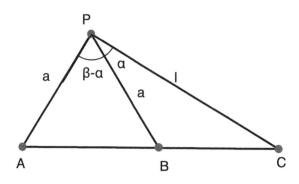

FIGURE 17.14

[P14.2] As shown in Fig.17.14: make an isosceles triangle $\triangle PAB$, P's vertex angle is $\beta - \alpha$. Let $PA = PB = $ a, and extend the base AB to C such that $\angle APC = \beta$, and then we have $\angle BPC = \alpha$. Obviously, we also have $\triangle APC \geq \triangle BPC$, that is:

$$al \sin \angle APC \geq al \sin \angle BPC$$

$$\therefore \quad \sin \beta \geq \sin \alpha$$

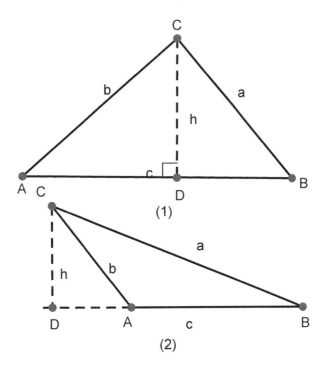

FIGURE 17.15

[P14.3] According to the usual definition, in the right triangle $\triangle ABC$, if c is the

hypotenuse then the cosine of acute angle $\angle A$ is $\quad cos\ A = \dfrac{b}{c}$. Then we define

$$cos\ \alpha = -cos(180° - \alpha)$$

which is basis of the proof.

As shown in Fig.17.15, suppose CD is one altitude of $\triangle ABC$, and $CD = h$. From Pythagorean Difference Theorem, when $\angle A$ is an acute angle, as shown in Fig.17.15(1), we have

$$\frac{b^2 + c^2 - a^2}{b^2 + AD^2 - h^2} = \frac{\triangle ABC}{\triangle ADC} = \frac{c}{AD}$$

From $b^2 = AD^2 + h^2$, we get

$$\frac{b^2 + c^2 - a^2}{2AD^2} = \frac{c}{AD}$$

$$\therefore \quad \frac{b^2 + c^2 - a^2}{bc} = \frac{2AD}{b} = 2\cos A$$

When $\angle A$ is an obtuse angle, as shown in Fig.17.15(2), from Pythagorean Difference theorem, we get

$$\frac{b^2 + c^2 - a^2}{b^2 + AD^2 - h^2} = -\frac{\triangle ABC}{\triangle ADC} = -\frac{c}{AD}$$

$$\therefore \quad \frac{b^2 + c^2 - a^2}{bc} = -2\cos(180° - \angle A)$$

$$= 2\cos A$$

[P14.4] Hint: (1) It's appropriate as long as we prove the case of

$0 \le \alpha \le \beta \le 90°$, since $\cos\alpha = -\cos(180° - \alpha)$.

(2) When α is an acute angle, then $\cos\alpha = \sin(90° - \alpha)$. We can use the property of sines to infer the property of cosines.

CHAPTER FIFTEEN

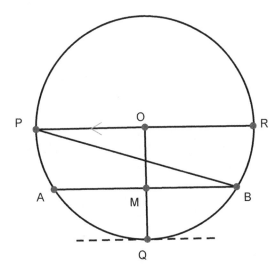

FIGURE 17.16

[P15.1] Not necessarily. You can take the following fact for reference:

As shown in Fig.17.16, OQ is the radius of $\odot O$, and M is the midpoint of the OQ. Draw a chord AB through M perpendicular to OQ and a diameter PR parallel with AB. Then we have $\triangle PAB = \triangle QAB$.

And we can prove the perimeter of $\triangle PAB$ is bigger than the perimeter of $\triangle QAB$. So we have

$$1 = \frac{\triangle PAB}{\triangle QAB} < \frac{PA + PB + AB}{QA + QB + AB}$$

Now, move the point P slightly upward such that the inequality

$$\frac{\triangle PAB}{\triangle QAB} < \frac{PA + PB + AB}{QA + QB + AB} \quad \text{still holds, yet} \quad \triangle PAB > \triangle QAB.$$

[P15.2] Hint: Apply proposition 15.1 .

[P15.3] Hint: Apply proposition 15.1. We can infer that the largest area of $\triangle PQR$ is

$\dfrac{2}{9} S_{ABCD}$.

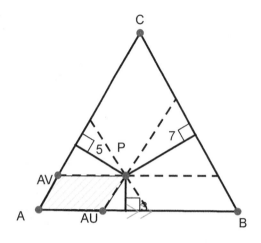

FIGURE 17.17

[P15.4] Hint: Draw three lines through *P* that are parallel with *AB, BC, CA* respectively. The dashed lines in Fig. 17.17 through P are parallel to the corresponding sides of the triangle, such that three parallelograms are formed inside the smaller triangles created by the parallel segments. Then the area of any triangle that is formed by cutting off a section along a dotted line is greater than or equal to twice the area of the corresponding parallelogram (the shaded part formed by the other two dashed lines). It is easy to calculate the altitude of $\triangle ABC$: 3+5+7=15, because the sum of the perpendicular distances from a point inside an equilateral triangle to its three sides is equal to the triangle's altitude. So, the area of the shaded parallelogram is $2 \times \frac{3}{15} \times \frac{5}{15} = \frac{2}{15}$. Then the area of the triangle that is cut off is at least $\frac{4}{15}$.

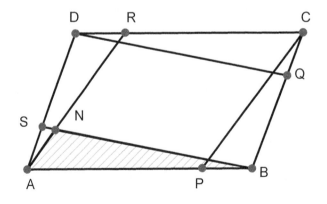

FIGURE 17.18

220

[P15.5] Hint: The key is to calculate the area of $\triangle ABN$, that is the filled area shown in Fig.17.18.

$$\frac{\triangle ABN}{S_{ABCD}} = \frac{\triangle ABN}{\triangle ABS} \cdot \frac{\triangle ABS}{\triangle ABD} \cdot \frac{\triangle ABD}{S_{ABCD}}$$

$$= \frac{BN}{BS} \cdot \frac{AS}{AD} \cdot \frac{1}{2} = \frac{\triangle ABR}{\triangle ABR + \triangle ABS} \cdot \frac{1}{3} \cdot \frac{1}{2}$$

$$= \frac{\triangle ABC}{\triangle ABC + \frac{1}{3} \triangle ARD} \cdot \frac{1}{6}$$

$$= \frac{\triangle ABC}{6(\triangle ABC + \frac{1}{9} \triangle ABC)} = \frac{3}{20}$$

Therefore, the area of required quadrilateral is $\frac{2}{5}$ of S_{ABCD}, that is

$$1 - \frac{3 \times 4}{20} = \frac{8}{20}$$

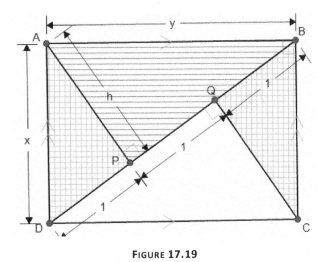

FIGURE 17.19

[P15.6] Hint: Suppose the width of a rectangle is $x = AD$, the length of rectangle is $y = AB$, and the distance from point A to the diagonal is $h = AP$ (as shown in Fig.17.19). The relationship of these three is

221

$$\begin{cases} x^2 + y^2 = 9 \ (Pythagorean \ Theorem) \\ xy = 3h \ (\triangle ABD = \triangle ABD) \\ y = hx \ (Co - Angle \ Theorem: \\ \qquad 2 = \dfrac{\triangle ABP}{\triangle PAD} = \dfrac{AB \cdot BP}{PA \cdot AD} = \dfrac{2y}{hx}) \end{cases}$$

We can get

$$x = \sqrt{3}, \qquad y = \sqrt{6}$$

$$\therefore \ xy \approx 4.2$$

$$\sqrt{18} = 3\sqrt{2} \approx 3 \times 1.414$$

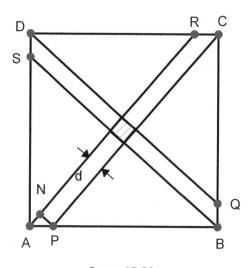

FIGURE 17.20

[P15.7] Hint: Use the method in exercise 15.5. We can also use the following method to calculate directly. As shown in Fig.17.20, the length of the center square's side is labeled *d*, and then

$$\frac{d}{AP} = \frac{AD}{AR} \ (\triangle PAN \sim \triangle ARD)$$

$$\therefore \ \frac{d}{\frac{1}{n}AB} = \frac{AB}{\sqrt{1 + \left(1 - \frac{1}{n}\right)^2} \cdot AB}$$

$$\therefore \quad d = \frac{AB}{\sqrt{n^2 + (n-1)^2}}$$

$$d^2 = \frac{AB^2}{n^2 + (n-1)^2}$$

From the given, $n^2 + (n-1)^2 = 1985$.

Solving for n, we have $n = 32$.

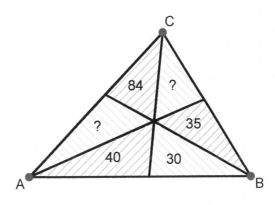

FIGURE 17.21

[P15.8] Hint: As shown in Fig.17.21, we can formulate the following equations:

$$\begin{cases} \dfrac{84+x}{35+y} = \dfrac{40}{30} = \dfrac{4}{3} & (1) \\[2mm] \dfrac{35+y}{30+40} = \dfrac{84}{x} & (2) \\[2mm] \dfrac{84+x}{40+30} = \dfrac{y}{35} & (3) \end{cases}$$

Among these three equations, we would select (1) and (3) to simultaneously solve:

$$\begin{cases} 3x - 4y = -112 \\ x - 2y = -84 \end{cases}$$

and get

$$\begin{cases} x = 56 \\ y = 70 \end{cases}$$

From this, we can calculate the area of the whole triangle, which is 315.

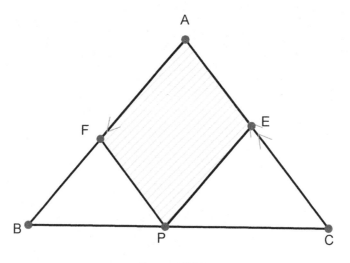

FIGURE 17.22

[P15.9] Hint: The problem should be considered under several different cases:

(1) if $\quad PB \geq \dfrac{2}{3}BC \quad$ then $\quad \triangle BPF \geq \dfrac{4}{9}\triangle ABC$

(2) if $\quad PC \geq \dfrac{2}{3}BC \ , \quad$ then $\quad \triangle PCE \geq \dfrac{4}{9}\triangle ABC$

(3) if $\quad PB < \dfrac{2}{3}BC \quad$ and $\quad PC < \dfrac{2}{3}BC \ , \quad$ let $\quad k = \dfrac{PB}{BC}.$

Then $\quad \dfrac{1}{3} < k < \dfrac{2}{3}.\quad$ Let $\quad \lambda = k - \dfrac{1}{3},\quad$ then $\quad 0 < \lambda < \dfrac{1}{3}.$

$$\therefore \quad \frac{\triangle AFE}{\triangle ABC} = k(1-k) = \left(\frac{1}{3}+\lambda\right)\left(\frac{2}{3}-\lambda\right)$$

$$= \frac{2}{9} + \lambda\left(\frac{1}{3}-\lambda\right) > \frac{2}{9}$$

$$\therefore \quad \frac{\square AFPE}{\triangle ABC} = \frac{2\triangle AFE}{\triangle ABC} > \frac{4}{9}$$